STRAIGHT FLUSH

ALSO BY BEN MEZRICH

Nonfiction

Sex on the Moon

The Accidental Billionaires

Rigged

Busting Vegas

Ugly Americans

Bringing Down the House

Fiction

The Carrier

Skeptic

Fertile Ground

Skin

Reaper

Threshold

STRAIGHT FLUSH

The True Story of Six College Friends Who Dealt Their Way to a Billion-Dollar Online Poker Empire—and How It All Came Crashing Down . . .

BEN MEZRICH

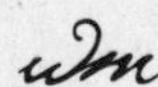

WILLIAM MORROW

An Imprint of HarperCollins*Publishers*

STRAIGHT FLUSH. Copyright © 2013 by Ben Mezrich. All rights reserved. Printed in the United States of America. No part of this book may be used or reproduced in any manner whatsoever without written permission except in the case of brief quotations embodied in critical articles and reviews. For information address HarperCollins Publishers, 10 East 53rd Street, New York, NY 10022.

HarperCollins books may be purchased for educational, business, or sales promotional use. For information please e-mail the Special Markets Department at SPsales@harpercollins.com.

A hardcover edition of this book was published in 2013 by William Morrow, an imprint of HarperCollins Publishers.

FIRST WILLIAM MORROW PAPERBACK EDITION PUBLISHED 2014.

Designed by Jamie Lynn Kerner

Library of Congress Cataloging-in-Publication Data has been applied for.

ISBN 978-0-06-224010-1

14 15 16 17 18 OV/RRD 10 9 8 7 6 5 4 3 2 1

For my dad, who inspired me to follow my dreams, and for Arya and Asher, who make me smile every single day

AUTHOR'S NOTE

Straight Flush is a dramatic narrative account based on multiple interviews, numerous sources, and thousands of pages of court documents. In some places, details of settings and descriptions have been changed to protect identities, and certain names, characterizations, and descriptions have been altered to protect privacy. In some instances I employ the technique of re-created dialogue, based on the recollections of interviewees, especially in scenes taking place more than a decade ago.

CHAPTER 1

DECEMBER 19, 2011

JUAN SANTAMARÍA INTERNATIONAL AIRPORT, SAN JOSÉ, COSTA RICA

Ten minutes before 5 A.M., a gray-on-gray sky was pregnant with the remnants of a passing storm, a thick canopy of clouds marred by occasional daggers of tropical blue and orange—and suddenly seven years disintegrated in a flash of reflected sunlight across the spinning glass of a revolving door.

Brent Beckley stepped through the threshold of the Central American country's main airport and into the poorly air-conditioned terminal. A little over six feet tall, with boyish features, a square jaw, and blondish-brown hair cut short over a wide, boxy forehead, Brent was moving fast, his five-hundred-dollar Italian-leather shoes clicking against the shiny linoleum floor. He was wearing a conservative dark blue suit with matching tie; there was a briefcase in his right hand and a heavy winter coat thrown over his left shoulder. Anyone looking his way might have assumed he was just another young, eager expat businessman on his way to an important meeting up north; business-clad

Americans strolling through Santamaría International were a common sight, symbolic of the expat community that had grown exponentially in the near decade since Brent had first arrived in the tropical country.

But the truth was, Brent Beckley was not on his way to a business meeting. In fact, he was quite possibly on his way to a jail cell. And the journey from where he'd started to where he was going was anything but common. He looked calm, cool, collected—shoulders back, head up—but on the inside he was terrified. He could feel the sweat running down the skin above his spine, and it required all his willpower to keep his knees from buckling, his body moving forward.

Ten feet from the blue-rope labyrinth that led through to Immigration and Security, Brent spotted a man strolling determinedly toward him and slowed his gait. At first glance, the man didn't look like a spy: thin, angular, with narrow cheeks, a sharp triangular nose, long legs lost in the folds of khaki pants, spindly arms jutting out past the cuffs of a white button-down shirt. The man was smiling, having recognized Brent immediately, though the two had never met. Brent tried to smile back, but the fear was playing havoc with the neurons that controlled the muscles of his face.

Brent was barely thirty years old, a small-town kid from backwoods Montana, a former frat boy who'd spent most of his adult life working for what he considered to be an Internet company; he'd certainly never expected to find himself rendezvousing in a tropical airport with a smiling spy.

Then again, the man wasn't *necessarily* a spy. From what Brent remembered from the letter he'd received the week before, detailing how the meeting would go down, the man's official title was some sort of "liaison" with the U.S. State Department, based out of the embassy in San José. And up close, even despite the sharp contours of his face, he looked much more like a kindly accountant than a menacing secret operative.

But if Brent had learned anything over the past seven years, it was that there were very few things in life that were *actually* black or white; most things tended to be a mix of both.

"Good morning, Mr. Beckley," the man said as he intercepted Brent a few feet from the entrance to the maze of blue rope. "My name is David Foster. It's nice to meet you."

Brent shook the man's hand, trying to think of a response. When none was forthcoming, Foster extended his other hand, offering two documents. The first was instantly familiar: Brent's U.S. passport—the same passport he had turned over to the State Department three days earlier. Glancing at the document, Brent felt his mouth go dry. He could see, even without looking closely, that someone had punched three holes through the center of the cover. Each dark circle tore at the pit of Brent's stomach. There was something so permanent and real about the sight of that passport; its mutilation seemed like such a malevolent and unnecessary act.

A week earlier, when Brent had first made the decision to turn himself in, the U.S. Embassy had requested a copy of his passport. Brent had been happy to accommodate, offering them the original document so they could copy it themselves; they had promptly confiscated it. Now he could see the result.

It seemed to be just another step in a deceptive game. Brent had already agreed to surrender, and he was in the process of moving his family to the United States—yet even that wasn't good enough.

Foster appeared to read Brent's thoughts and quickly shifted the invalidated passport to the side, revealing the second document in his hand: a thin, similar-looking passport, this one with its cover still intact. Brent took both documents from the man, inspecting the second, smaller booklet—and saw that it was dated for a single day's use. Brent was still free to travel like any other American citizen—*for the next twenty-four hours.*

There was a moment of awkward silence, and then Brent finally shrugged, shoving the two passports into his suit pocket.

"What now?" he asked.

Foster's expression turned soft, and he jerked his head toward the blue ropes behind him.

"We've got an hour to kill before your flight. You want to get a cup of coffee?"

It wasn't quite what Brent had expected—but again, none of this could have been anticipated. He nodded and followed the thin man toward Immigration.

It was the fastest Brent had ever moved through the Costa Rican airport; usually, security took forever, especially for young Americans like him. In Brent's experience, some of the native immigration officers seemed to take a special pleasure in hassling young American men traveling to and from the States. Brent assumed it had to do with the massive inequities between the two cultures; to the average Costa Rican, Americans were rich, entitled, and usually obnoxious. From what Brent had seen of the mobs of northerners who kept the local tourist economy alive—usually large groups of men who spent mornings splayed out across the pristine beaches like bleating, bloated, bleached, and beached marine animals, and evenings carousing through the legal brothels that put red-light districts around the world to shame—well, maybe the immigration officers weren't that far off. At the moment, Brent could only marvel as he was towed through Immigration and Security at a near-Olympic pace; Foster seemed to know everyone who worked at the airport, and even more helpful, the man's Spanish was impeccable. He spoke like a native—though from what Brent could piece together, it appeared that Costa Rica was just one stop on a colorful, government-sponsored road trip that had extended from a military academy in Virginia, through a five-year stint in Iraq, to a half dozen embassies across South and Central America. Even if Foster wasn't a spy, he'd certainly lived like one. Yet by the time Brent lowered himself onto a stool in a quiet corner of a dingy coffee shop—just beyond the last security checkpoint before the waiting area for Continental Airlines, the carrier that would take him out of his adopted home, possibly forever—he felt as comfortable with the man as one could possibly be, under the circumstances. Foster wasn't a bad guy, and he wasn't the enemy. He just worked for them.

Foster ordered for both of them, making small talk as the uniformed waitress brought them Styrofoam cups filled with tar-black coffee. The first sip put strength into Brent's knees and warmed his throat enough to make the words come a little easier.

"This is just so crazy," he said, the most words he'd strung together since he'd stepped into the airport. "I'm not even sure what I'm doing here."

Foster smiled, sipping his coffee. "Getting on a plane to New Jersey."

Brent must have given him a look, because Foster laughed.

"Kid, it really does help to keep things simple in your head. Take it one step at a time. Right now, you're drinking a shit cup of coffee in a shit coffee shop. An hour from now, you'll be boarding a 737 to Newark. Real simple, like that."

Brent nodded. The guy was probably right. Keep his thoughts simple, keep focused on the moment, the little picture—because when he let his mind go after the big picture, well, things got really dark and confusing.

"It just doesn't seem fair."

Foster shrugged. "To tell you the truth, I don't understand why they want you either. But that's not my job."

It was good to hear, but Brent couldn't help finishing the man's thought: Foster's job wasn't to understand why Brent was being prosecuted; it was to facilitate the situation. Or more bluntly, make sure Brent got on that airplane. Brent couldn't help wondering what Foster would do if he suddenly changed his mind—just turned and headed for the airport exit. Would Foster try to stop him?

Brent immediately chided himself. He was letting his fear get to him. He'd already made the decision. The wheels were in motion.

But still.

"I'll probably get some points for surrendering. I mean, I could just stay here in Costa Rica, right?"

He'd spoken to enough lawyers to know that technically, for the moment anyway, he was correct. One of the key points for extradition was that the crime you were accused of committing had to be illegal in both jurisdictions. As far as he—or his lawyers—could tell, what he'd done, what he was accused of doing, was legal in Costa Rica. Hell, it was legal pretty much everywhere in the world—except for the United States. And even there—well, he and his legal team still weren't entirely clear.

"Maybe," Foster agreed, shrugging his shoulders. "I mean, we probably couldn't have extradited you. But that doesn't mean we couldn't get you."

Brent looked at him. There was a glint in Foster's eyes as the thin "liaison" leaned close, over the table.

"When we really want somebody, we work with our friends, in whatever country we happen to be. A few phone calls, a little back-and-forth, tit for tat. We get them to cancel your immigration status, and next thing you know, you're being deported. Guess where?"

Foster was still smiling, but his thin features didn't seem quite as amiable as before. Brent stifled a shiver.

"Put a bag over my head, hit me with a truncheon, shove me into the trunk of a car?"

Foster laughed. "Come on, kid. You've been in Central America too long. This is the U.S. government you're talking about. We're civilized."

Brent pretended to ease back against his stool, but his muscles were tense, his nerves once again feeding rubber into his knees. When the U.S. government wanted to lock someone up, they didn't need black bags, truncheons, and trunks of cars. They simply passed a law to make whatever their target was doing illegal. Then they punched holes in his passport.

Brent exhaled, taking a deep drink from his coffee.

"So I guess I'm doing the right thing. It's just . . . well, this wasn't how this was supposed to have gone down."

Foster shrugged again. He'd heard the line before, probably many times. The thing was, in Brent's case, it was more than a cliché. Seven years earlier, when he'd strolled through this very airport for the first time—a kid barely out of college, on his way to join four of his best friends chasing a dream that at the time seemed so real and possible—it had felt like the beginning of a grand, exotic adventure. And in many ways, those seven years had been just that—grand, exotic, exciting, and at times unbelievably profitable. Brent and his friends had built something amazing.

And then, just like that, in a flash as quick and blinding as sunlight on a glass pane, it had all come crashing down.

"Yeah," Brent said, and sighed, crumpling his now empty Styrofoam cup in the palm of his hand, "maybe we were stupid, but none of us pictured it ending like this."

Two hours later, Brent toyed with the recline lever of his first-class aislc scat, trying and failing to find a setting that might relieve the dull ache that had settled into his bones once the narrow-bodied Continental 737 had reached its cruising altitude. He knew his efforts were futile; his discomfort had nothing to do with the seat, or the fact that even in first class, his legs were pretzeled together. His body hurt because now that he was alone in the confines of the airplane, his mind couldn't help whirling forward, to what was coming. And even at his most optimistic, Brent knew that it was going to be one hell of a hard landing.

He desperately wanted a drink, but alcohol would be a bad idea. His head needed to be clear. Even his seat was a source of mild anxiety—he had purchased a ticket in coach, but someone had shifted him to first class, front row, aisle. He wasn't sure if federal agents were going to get on the plane and take him off in handcuffs or if they'd let him walk through the Jetway under his own power. Either way, the gnawing thought of what was awaiting him would make this the longest flight of his life.

As the plane began to jerk and jag through a spot of mild turbulence, Brent shut his eyes, forcing his head back against the faux-leather headrest. Eyes closed, he was not surprised to immediately picture his wife and two young sons; at that moment they were probably beginning the process of setting up residency in Salt Lake City, where he planned to eventually join them. That little family was, without question, the most treasured part of his life. They were the reason he was on that 737. The reason he'd surrendered—even though in the minds of some of his friends, surrendering was akin to giving up without a fight.

The bottom line was, Brent's wife was Colombian, his kids Costa Rican; if he was going to have any chance of giving them a life in the United States, of having his kids become full citizens like their father—he was going to have to make a deal.

And in a way, that had made his decision easier. There had been other options—and not just staying in Costa Rica. His older brother—stepbrother, technically, whom Brent idolized and respected more than anyone else on earth—had gone a very different route. Scott didn't like to use the word *fugitive* because, in truth, he wasn't running, nor was he exactly hiding—the U.S. government just couldn't get him as long as he stayed within the borders of the tiny Caribbean island he now called home. But for Brent, returning to the States had always been the endgame; the government's offer to assist in his family's relocation had tipped the scales, and despite the anxiety Brent felt about his own future, at least for his family, he was pretty sure he was doing the right thing.

For the moment, he did his best to cling to that minor solace. He had to believe that whatever they did to him, he had made the best decision for his family. He kept his eyes closed, that thought firmly in place, until the plane finally began its descent into Newark.

It wasn't until he heard the quiet rumble of the Jetway moving into place that he finally opened his eyes. He watched the flight attendant going to work on the door; a few clicks and a grunt later, the attendant stepped back, revealing the orange-lit tunnel stretching forward into the depths of Newark International Airport.

Brent gave it a full thirty seconds before he decided it was okay for him to be just another passenger, at least for a little while longer. He retrieved his briefcase and overcoat, then headed for the Jetway.

It wasn't until he'd reached the end of the long, angled tunnel that he saw the immigration officers. He quickly counted six of them, all in uniform—and every one armed. Nobody had a gun drawn, but even so, the sight of those leather holsters, pitched high on each officer's hip—it was enough to take Brent's breath away. He did his best not to stumble as he made it the last few steps to the end of the Jetway.

The closest officer held up a hand, palm out.

"Are you Brent Beckley?"

Brent nodded. The man turned his hand over.

"Passport, please."

Brent fumbled with his coat for a second, then retrieved the single-day passport and gave it to the officer. The officer checked it, showed it to one of his colleagues, and then all six moved forward, taking positions around Brent. The lead officer gestured with his head—and suddenly they were moving forward through the terminal in what appeared to be a diamond formation, with Brent right in the middle.

Christ. It was the most absurd feeling. The officers were walking fast, and Brent was nearly skipping to keep up. People stared as they went past—pointing, whispering, a few even snapping cell phone pictures.

The mobile diamond advanced unimpeded through Customs and out into the main baggage claim area. On the other side of baggage claim, the officers finally broke formation, and Brent was handed off to two middle-aged men in white shirts and dark ties. One of the men showed Brent an FBI badge, the other a badge marked HOMELAND SECURITY. The officers were exceedingly polite, but by this point Brent's heart was pounding so hard, he could barely understand what they were saying. They walked him out of the baggage area toward the terminal exit.

Frigid air splashed against Brent's cheeks as they stepped outside onto the sidewalk, shaking some of the fog out from behind his eyes. He immediately saw a vaguely familiar face, a woman in a dark suit hurrying toward him from the curb, a forced smile on her lips. Brent recognized her as one of the low-level associates from the law firm he'd hired to handle his criminal proceedings. While Brent stood between the officers, she retrieved his briefcase and overcoat. Then the FBI agent pointed at his watch.

"Better take that too. And his belt, and cuff links."

Brent swallowed, then slowly went to work on the watch. He suddenly noticed that his fingers were shaking, and it took a good minute to get from the watch to the cuff links. His belt was a little easier, though his pants felt strange without it; luckily, he'd put on an extra pound or two in the anxiety-filled weeks leading up to his surrender.

After he handed over the items, the FBI agent reached into his back pocket. Out came the handcuffs, like a fist to Brent's gut. When the cold metal touched his wrists, then closed—tight, too damn tight—Brent fought the urge to break down. It all seemed so goddamn unfair.

But instead of complaining, Brent didn't say a word. He let the officers lead him to a waiting black sedan. The Homeland Security officer got behind the wheel; the FBI agent slid into the back next to Brent. A moment later they were off, tires rolling against pavement, winding their way out of the airport and onto the Jersey Turnpike.

Brent tried to find a comfortable position, but the tight handcuffs made it nearly impossible. Instead, he tried to concentrate on the sound of his own breathing. His chest felt constricted, his mouth dry as cotton. He felt himself losing all sense of time as the gray turnpike flickered by outside the tinted window to his left. Was it still morning? Afternoon? How long had they been driving? Were they in New York, or still in New Jersey?

Eventually the silence began to get to him, and he quietly cleared his throat.

"So, are you guys just here to process me today? Or have you been working on my case for a while?"

The agents shared a look in the rearview mirror. Then the FBI agent grinned.

"We've been onto you for a long, long time, Mr. Beckley."

Brent forced a smile of his own.

"Well then, I guess it's nice to finally meet you."

As he turned back toward the window, the sight of something in the distance made him blink. Tall, rising out of a faraway mist, reaching toward the sky: the Statue of Liberty. Brent was seeing it for the first time. *Handcuffed, sitting next to an FBI agent.* He felt the haze of unreality coming back. Once again, he lost all sense of time.

The next hour went by in flashes. An FBI processing center, somewhere in midtown Manhattan—they'd driven in through a gated basement entrance, then gone up in an armored elevator to a cubicle-filled office full of printers, copy machines, and many more agents in white shirts and dark ties. Fingerprinted, photographed, then back into the elevator, returned to the sedan—and on to another faceless building, another gated basement entrance. At that point, the two officers handed him over to a pair of U.S. marshals, who took him into a similar elevator. The marshals were decidedly less polite than the two previous agents; they were large, burly men, with crew cuts and matching cruel grins. When one of them noticed Brent's expensive shoes, he pointed a thick finger at the silver clasps.

"I'm gonna need to rip these off," he said. And a second later the marshal was on his knees, yanking at the clasps with his meaty paws. After a few minutes of grunting and groaning—while Brent did his best to keep from toppling over—he eventually gave up.

Then they were in another processing center—more fingerprints, more photographs. Brent was handed off to different officers and eventually led via a tunnel to another building. Finally, nearly eight hours after he'd taken off from Costa Rica, he arrived in a jail cell.

Barely larger than ten by ten, it had a low ceiling, white walls, a pair of steel benches suspended beneath a tiny barred window. There was a scruffy-looking man sleeping on one of the benches; as the barred door slammed shut behind Brent, the man momentarily looked up before going right back to sleep. Brent moved a few feet into the cell, then just stood there, staring at the walls, the window, the bars. Everywhere he looked, he saw rivets, some of them rusted, some of them shiny. Rivets, thousands of rivets, running up the corners of the walls, around the window, along the door. So many goddamn rivets.

Brent felt his shoulders begin to sag.

He truly hoped that he was doing the right thing. Because it was suddenly very obvious: he wasn't getting out of that cell until somebody came and let him out. It was maybe two in the afternoon; he had a whole day ahead of him. He was barely thirty years old; he had a whole life ahead of him.

It wasn't supposed to end like this.

It wasn't supposed to have ended at all.

In the beginning, it had been something so special, so wild and cool—and simple. A group of best friends and two brothers, who had set out to do something different.

None of them could have ever imagined how quickly something so simple could become something huge—or how equally quickly it could all come crashing down. They had risen so far—Christ, at one point, they had been days away from being billionaires.

Now Brent was counting rivets in a prison cell, his brother had sequestered himself on an island the size of a Minnesota shopping mall, and the others had scattered all over the world, facing futures as uncertain as his own.

No, Brent thought to himself as he once again shut his eyes, picturing his wife and his two little boys.

This isn't how it was supposed to go down at all . . .

CHAPTER 2

SEPTEMBER 1997

"Here they come, boys. Give me your tired, your hungry, and your wretched. Especially your wretched. Some of my best friends are freaking wretched."

Garin Gustafson grinned as he rose from the front stoop of the SAE fraternity house and tossed a half-empty beer can over his left shoulder. The can arced upward like a Scud missile, hung in the air for a full beat, then spiraled down in a flash of spinning aluminum. Pete Barovich and Shane Blackford, seated two steps up the dilapidated front porch of the aging frat house, cursed as they ducked in tandem. The can hit the edge of the step behind them, then pinwheeled back into the air, spraying beer as it went.

"The only thing wretched is your aim," Pete said, coughing. "It's no wonder you've stayed with the same girl since high school. You can't hit the side of a barn with a tin can. What chance you got making a college girl smile in the dark?"

Garin held up a middle finger without turning around.

"If I'd been aiming for the barn, you'd be picking shingles from the roof out of your hair. I was aiming for the two chickenshits on the front stoop. I'm hearing a lot of squawking, so I couldn't have been too far off."

Garin raised his arms above his head, stretching his spine, as his two friends wrung beer out of their Sigma Alpha Epsilon sweatshirts. *Goofballs.* And they had the temerity to question his aim. He had taken the house to how many intramural hoops wins in his two years with the fraternity? He could have put the beer can right through the second-floor window above the porch, set it down smack in the center of the desk in Shane's nearly OCD-level, immaculately clean room.

Not that Garin wanted to put another hole in the sagging, multistory monstrosity they called home. As much as it looked like a pretty white barn, all gussied up for the first night of the University of Montana's infamous Greek Week, structurally, the frat house was as rotten as a cow with intestinal worms. From the outside, it was something right out of *Pleasantville,* fitting in perfectly with the dozen other frat and sorority houses that lined the bucolic stretch of suburban Missoula, kitty-corner to the main campus. But if you stuck your head past the door frame, craned your neck just a little bit—well, it was a different story. Loose wires hanging from sparking sockets, stairs that disintegrated beneath your feet, ceilings that drizzled down plaster, toilets that backed up sewage every seventh night like it was some sort of Sabbath ritual—one more beer can through a window might just bring the whole goddamn thing down, and how the hell would that look? Garin's first Greek Week as house rush chair, and him responsible for the demolition of the SAE house? While Pete, SAE president, and Shane, maybe the most liked member of their class, looked on?

That wouldn't do at all. Garin would never intentionally do anything to harm his beloved SAE. He could still remember the first time he stepped into that house after he'd survived its notorious Hell Week and stood shoulder-to-shoulder with the guys who would become his brothers overnight. Hell, back then he'd been straight out of farm country, plucked from a small town in the middle of nowhere called Conrad, way up near the Canadian border. A place where life revolved around high school sports, farming, and cows. He'd been a wide-eyed kid, shocked by nearly everything he saw. To him, Missoula had been a damn big city.

He'd changed a lot since then; he was more confident, better with the girls. He'd kept his six-pack and his farmer's tan, but he liked to think he wasn't nearly as naive. He certainly wasn't the same bumpkin who'd sold a cow to buy his first car—a rear-wheel-drive Mustang that was completely pointless in the Montana winters. But he was still, at heart, 100 percent country.

He wasn't social chair because he could play basketball. He was the one they sent out onto the street during Rush Week because inside, a part of him would always be that wide-eyed, small-town kid.

"Well, you'd better get cracking," Shane chimed in. "Find us some new pledges. Someone's got to pay for the upkeep of our coop. Us chickenshits get restless, and when we get restless, things get broken."

Garin jibed left, even before he heard the can whistle toward his back. It spiraled harmlessly past his shoulder, then skipped across the loosely paved street that ran between the frat and sorority houses. He laughed, then followed in the can's wake, finally taking up position a few feet past the edge of the curb, facing the campus.

It really was a beautiful sight. The whole block was congested with college kids, most moving in large groups, some in twos and threes. Everyone seemed to be coming down the center of the tree-lined street, which had been closed off to car traffic for the occasion. On either side, beyond the mailboxes, manicured hedges, and the odd pickup truck sporting bales of hay shaped into a patchwork of Greek letters, the houses were alive with everything that made this place the center of social life at the state's biggest university. Parties were already sliding toward full swing at nearly every other house on the block, even though it was barely 6 P.M. and still reasonably light outside. But SAE liked to wait a bit before tapping the multiple kegs lined up behind the bar in the basement. Pete liked to say, you have to let the night breathe a bit, like a fine wine, before you started popping corks. From Garin's vantage point on the street, it looked like corks were flying all over Greek Row. But he wasn't worried. SAE had a reputation that drew a respectable crop of pledges every season. By 10 P.M., he knew, the house behind him would be throbbing with good music, foaming with decent beer, and, most important of all, teeming with pretty girls. And girls, of course, were the currency that held the whole system together.

At that very moment, he knew, two stories above his head, a trio of sorority sisters were taking their seats on the deck that jutted out above the SAE front porch. All night long, the girls would smile and wave at the groups of guys wandering by—a little tease of the night to come. Girlfriends of three of the fraternity brothers, the girls gave the place a bit of a Mardi Gras feel. Of course, the balcony babes had been Pete's idea. Garin had to give him that: he was one hell of a marketing genius.

And already the girls were working their magic, as the first few groups of guys passed by. The girls laughed and waved, and the freshmen ate it up. Garin had to smile—he understood those freshmen; he knew them well. They were all pure Montana, from Billings, Missoula, and the thousands of smaller towns scattered across the wide farm and ranching state. Mostly lower-middle-class kids from squat in the middle of a depleted country economy—not on food stamps, but not rich either. Maybe their parents made forty thousand a year. They played sports, they drank beer, they liked cars, and they loved girls.

Garin let the first group move by without stopping any of them, then did the same with a second. He wasn't being picky; they were all good prospects. The truth was, this particular year it would mostly be a numbers game. The house was falling apart, so they needed lots of new blood. Basically, a minimum number just to pay the bills. But Garin didn't want to start his first recruiting night with just anyone. He was a born athlete, had been playing sports at a near-elite level since before he could read, and he knew what it meant to build a winning team. His job as social chair was to put together a pledge class that would last, as brothers, for a lifetime. And every athlete in the world knew that a good team began with a strong anchor. Garin intended to start his night by finding that anchor.

So he did his best to ignore Pete and Shane, who were shouting epithets in his direction—a mix of idiotic insults taking aim at everything from his elongated physique to his small-town origins. No matter how hard they pushed him, he was going to do this right.

And then, right in front of him, not four feet away, moving down the center of the street at the tail end of a group of guys—this kid was definitely something different. Big eyes, intensely green, flicking back and forth as he took in everything around him, like it was all just a show put on for his benefit. Chiseled features under longish locks of auburn hair. Good-looking but not effeminate, obviously an athlete of some sort, though not as tall as Garin or anywhere near as ripped. At the same time, there was certainly something off about the kid, especially in the way he was dressed.

He was wearing a black leather jacket that was at least a size too small for him, the sleeves barely reaching halfway down his forearms, over a white polo shirt with the collar popped up, partially obscuring what looked to be a thin silver necklace. Below the jacket, faded dark jeans, torn and scuffed around the knees, but not in the stylistic way that you might pay double for at a department store—torn because at some point those jeans had seen him through some sort of trouble. And then, below the jeans, brown boat shoes with no socks.

Who the hell dressed like that? And yet, though the kid looked absurd, he didn't seem insecure at all. In fact, he appeared positively cocky, grinning, puffing his chest out, cracking jokes to anyone who would listen. The kid made a damn good first impression.

Garin waited until his target was just a few feet away before making his move. He stepped forward, effectively blocking the street, and offered a hand along with his most gregarious smile. At six three he towered over many freshmen, but he was lean enough not to be threatening. Not that this kid looked like he could be easily intimidated. He nearly crushed Garin's hand with his own.

"Garin Gustafson," Garin said. "Rush chair, SAE. How's your night shaping up?"

"Got some idea how it's gonna start, no idea how it's gonna end. So I say it's shaping up pretty damn good. Scott Tom. Nice to meet you."

Garin laughed. "I'm sure you've heard about us, so I'm not going to bore you with some clipboard crap about diversity, history, or school spirit. SAE is a great bunch of guys who like to have a whole lot of fun."

Scott pointed toward the girls on the balcony.

"Is there more where they came from?"

"Right to the point," Garin said. "Save the small talk for the ladies—I like that. Come back in about four hours. The party should be in full swing by then. I don't think you'll be disappointed."

There was a flash of mischief in the freshman's eyes. "I just might take you up on that. And I hope you're right—you wouldn't like me when I'm disappointed."

Garin couldn't tell if the kid was joking. It was such a strange thing to say. Before Garin could respond, the kid was already moving away, catching up to the group he'd come with. Garin made a quick decision and took a step after him.

"Hey, if you're thinking of heading downtown to pregame, maybe this will help you out."

Garin fished in his shirt pocket and pulled out a plastic card. It was a Montana driver's license that he'd found floating at the edge of the river when he was out tubing with a beer cooler earlier that week. Most likely a fake ID that someone had discarded after a drunken night—maybe the original owner hadn't had the confidence to pull off the role of a twenty-four-year-old organ donor from Billings. Garin was pretty sure Scott would not have that problem.

He tossed the license in Scott's direction. Scott picked it out of the air with a flick of his wrist, looked at the picture and the date of birth, then gave Garin a thumbs-up.

"Cool. Thanks, man. And I thought I was going to have to get by on my natural charm."

With that, the kid turned and headed back on down the road, leaving Garin wondering if the moment had been the start of something interesting—or if he'd never see that strange, confident green-eyed kid again.

CHAPTER 3

Fifty-seven minutes. I shit you not. Paid off half the house debt and got myself banned from the Missoula County Fairgrounds for three years. But it was a hell of a party."

Pete Barovich crossed his arms against his chest as he leaned back against an aging mahogany upright piano, shoulder-to-shoulder with a brunette in a midriff-baring halter top and low-rise skinny jeans. The girl had to be a volleyball player, she was so damn tall. She had a few inches on Pete, even without her three-inch chunky white heels, and Pete was seriously concerned that the two of them were testing the structural integrity of the antique instrument behind them. After all, the piano had already lost most of its keys, and two of its legs were so chipped and worn they looked like they'd been gnawed upon by a giant rodent.

Certainly, the thing hadn't worked as a musical instrument since Pete had entered the house—which meant it fit right in with the rest of the vast living room's décor. The place was actively falling apart, and now that the inaugural rush party was in full swing, even the floor beneath Pete's feet seemed to be bucking and swaying in rhythm with a few hundred drunk college kids rapidly being whipped into an alcohol-lubricated frenzy.

"A thousand goldfish in fifty-seven minutes?" The brunette gasped. "That's disgusting."

She covered her mouth, feigning nausea. But Pete could see from the look in her eyes that she was actually impressed. He was pretty certain she'd heard the story before; it had fast become legend across the University of Montana campus and was one of the reasons Pete had been elected president of the fraternity while only a junior. Throwing a party like tonight's was one thing, but throwing a party that would be talked about for years to come was an accomplishment that made it onto your résumé.

The Goat Barn Party—as it had become known—had taken place a year earlier, shortly after Pete had been elected SAE's social chair. The house had been in desperate need of money; a cocaine scandal the year before had nearly gotten the frat kicked off campus and had put the place six figures into debt. Pete had decided the only way to save the house was to party their way out of debt. He had rented out a nearby state fairground, known for its working goat barn, and arranged the delivery of forty kegs of beer, a live band, and plenty of publicity. The goal was to get a ton of kids to attend—and charge them all—but in a place like Montana, that was easier said than done.

So Pete had come up with a unique plan. He'd gone to the local pet store and purchased five thousand goldfish. Anyone attending the party had a choice. Either pay full price—five dollars a head—or swallow a goldfish and get in for one dollar. Girls who swallowed a goldfish got in for free.

"Talk about setting the mood—swallow a goldfish at the front door, and by the time anyone set foot in the goat barn to hear the band and drink some beer, their inhibitions were gone. By the time it was over, I think nearly every cop in the city had made an appearance. The only reason I didn't spend the night in jail was at least half the police force went to my high school."

The girl laughed, wriggling a little closer. Pete felt the warmth of her bare shoulder against his arm and immediately bumped her up in his mental rankings. Garin wasn't the only one assembling prospects that night.

Sliding a hand around her waist, Pete glanced out across the crowded living room. He tried not to grimace as he scanned the pathetic attempts at furniture: a handful of ratty couches strewn about the scuffed hardwood floor, the sofas upholstered in a clash of haphazard colors, pillows stained so deeply it was impossible to imagine what the original shades might have been. Shelving units that looked like they'd been plucked out of the trash were cluttered with old books, beer cans, and various trinkets that seemed collegiate—cloudy glass steins, broken sports trophies, old college yearbooks. The walls, where they were visible between the shelves, were cracked and peeling; the worst cracks were vaguely covered by oil paintings purchased at various garage sales and flea markets, depicting everything from sailboats to farm animals. No dogs playing poker, though a bit of framed velvet would have classed up the place enormously. Everything looked old while somehow avoiding any pretense of gravitas.

And yet Pete and his brothers loved that house. From the looks of the raucous crowd filling every inch of the living room, the feeling was infectious, at least enough to have attracted a good assemblage of the freshman class.

Pete momentarily forgot about the brunette leaning next to him as he surveyed the new talent Garin had invited into their carnival. A few faces he recognized from one or another of the various sports teams, whose recruitment had often included a tour of the houses on Greek Row. Others he recognized from the street outside. As a group, they seemed to be having a good time, reveling in the abundance of free alcohol, ear-shattering music, and the few dozen sorority girls whose main function was to draw the attention away from the state of the house itself.

Yes, if Pete said so himself, he sure was a marketing genius. He caught sight of Shane standing between a pair of girls in matching jeans shorts, both wearing shirts that may as well have been bikini tops—and raised an eyebrow. The prospects seemed good indeed.

And then Pete's gaze settled on another recognizable face, a few feet behind Shane, in a corner of the room between an overturned loveseat and an oversize plaster bust of Gary Cooper that one of the brothers had won in a card game.

Scott Tom.

Even if Garin hadn't taken a special interest in the kid earlier that evening, Pete would have given him a second look, and probably even a third. He was still wearing that ridiculous leather jacket, the cuffs inches above his wrists, as well as those damn boat shoes. And his eyes were still the same pools of green, taking in everything at once, constantly searching, scanning—like he was mentally disassembling all the furniture and putting it back together in a way that made more sense. This kid was different.

"Hold that thought for a moment," Pete said to the brunette next to him, who now had a hand running up his right thigh. "I've got some house business to attend to."

The girl followed his eyes, and then a look crossed her face.

"Not him. Anybody but him."

Pete looked at her. "You know him?"

"We all know him. And not just my sorority. Go ahead, ask around."

"Really? It's only his first semester. How much damage could he have done this quickly?"

"Two rooming groups, three girls each, that I know of in my house alone. Slept with one roommate, then the next, then the next—all in the same weekend. And I heard that he's already been banned by both Sigma Tau and Pi Theta E."

Pete whistled low. *A new town slut.* The brunette's warning was having the opposite effect on him than she'd intended. Pete was even more intrigued by the kid. He himself had built up quite a reputation, beginning back in high school—when he'd been known as Porno Pete, first, because of an uncanny resemblance to a famous porn star with the same first name, and second, because he'd racked up pretty good numbers in his school district, even before he'd won a couple of wrestling trophies.

Pete patted the girl's hand, then delicately lifted it off of his thigh.

"Sounds like I'll need to be careful with this one. Don't worry, I'll make sure he's on a strict leash tonight."

The girl rolled her eyes as Pete pushed off the piano and strolled across the room toward the far corner. Along the way, he was shaking hands, slapping backs, pausing a few times to share a beer. Eventually he wound his way past Shane, admiring the two bikini girls, then made his approach. Scott was still standing alone when he got there, but if the kid felt insecure in any way, he certainly didn't show it.

His handshake was firm, his smile strong.

"This certainly looks like the place," he said, in way of a greeting. Then he gestured toward the brunette who was still against the piano, leaning back to reveal even more of her criminally flat stomach. "I think her name is Julie, right? Phi Beta? They serve a great brunch on Sundays. Maybe I'll run into you there, one of these weekends."

Pete laughed. This kid was not going to disappoint. And from the way he was looking around the room, noticing every girl, it seemed clear where his priorities lay. Fair enough; most guys joined a fraternity for the girls. Brotherhood usually came as a surprise.

"Pete Barovich," Pete said, introducing himself. "I'm from Billings. Basketball, football, wrestling, and tennis. Maybe we played against each other at some point?"

Scott shook his head. "You're big city. I'm one hundred percent trailer park. The only way you might have met me in high school is if your mom worked for child services."

Pete blinked. It took him a minute to realize the kid was serious. If he thought Billings, Montana, was a big city, he probably really had been brought up in a trailer. Which could be good or bad. Garin had stepped right off the farm, and he was now one of Pete's best friends. Though a kid from a trailer park might not help that much with the house's bottom line. It was the annual dues that kept the place afloat, after all.

"Dad's an insurance salesman," Pete responded. "Mom is a nurse. But I like to say, doesn't matter where you start—"

"Only matters where you end up. Better yet—who you end up with."

Scott grinned, jerking his head toward the brunette who was still crawling up the piano. Then he gave Pete a punch in the shoulder.

"Don't worry, I keep my scars on the inside. Except the ones you can't really hide, like the cigarette burns and the razor wounds."

Pete opened his mouth, but Scott waved him off.

"Kidding. My dad got me out of that hellhole before I turned into a sociopath. He's an investment banker in Seattle. If you let me into your house, I'm sure you'll meet him. He's almost as good at getting kicked out of sorority houses as I am."

Pete laughed. Talking to Scott was like riding upside down on a Ferris wheel—you had to keep your hands on the fucking safety bar. He gave Scott's shoulder a squeeze, feeling the thick leather of that distressed jacket.

"I got a good sense about you. If the rest of the guys like you as much as I do, and you somehow survive Hell Week, I think you could end up giving me a run for my money with the girls. And I sure as hell like a challenge."

"Then cut out the chitchat and get me a motherfucking beer. The night is just getting started."

Pete caught sight of Shane again, over Scott's shoulder. Shane gave him a thumbs-up; there wasn't going to be much debate about this one.

Maybe Scott Tom had grown up hard, but Pete had a feeling the kid was going to fit in well with the SAE brothers. Hell, with a smile like that, and his almost obscene level of confidence, maybe one day he'd be running the whole goddamn house.

CHAPTER 4

So this is how it ends, Scott Tom thought as he struggled to disentangle himself from the heavy blanket that someone had tucked much too tightly around the corners of the queen-size mattress in the center of his room. For a brief second he pondered who might have been responsible for the blanket, because he sure as hell hadn't tucked it in himself—but then there was that sound again, an ear-shattering roar that seemed to split the very air, and the whole house was suddenly trembling around him.

If it really was an earthquake, he didn't want to die like this, half-naked and trapped under a blanket. Even worse, down by his feet he could feel what seemed to be a sequined tube top. And up by his elbow, a pair of jeans, with a pair of thong underwear still inside.

Kicking as hard as he could, he finally got himself free of the blanket. The roar went off a third time, nearly knocking him onto the floor. He pushed himself to his feet, using the shiny chrome stripper pole he'd installed next to his bed for leverage. He'd always known the pole would come in handy, despite what Pete and the rest of the brothers thought of his home-decorating tastes. A stripper pole just made sense—especially since he'd painted the walls a searing bright red and installed those mirrored panels along the ceiling. But then again, what the hell did he care what his frat brothers thought of his room's décor? In the semester and a half since he'd moved into the house, he'd discovered that he was by far the most imaginative of the group.

Using the pole as an axis, he flung himself toward the door. Just as his hand reached the knob, he heard another cacophonous burst.

But now that he was closer to the source, it no longer sounded like an earthquake. Flinging the door open, he found his suspicions confirmed—though his initial confusion and thoughts of impending doom were understandable.

Scott had seen plenty of crazy things since he'd moved into the SAE house, but the sight before him was pure bedlam. Half a dozen brothers were out of their rooms, lining both sides of the third-floor hallway, most in similar states of undress. And there, like an untamed beast pawing through the hall carpet and into the very floorboards, some idiot on a four-hundred-pound Harley was in the process of taking out half the banister as he spun the massive steel-and-chrome motorcycle in a wheelie. With a warrior's cry, the guy suddenly headed back down the stairs, carpet and wood splinters spraying in his wake.

Christ. At least the idiot was wearing a helmet. Scott shook his head, rubbing the last vestiges of sleep out of his eyes. He was about to go back into his room to try to solve the mystery of the tucked-in covers and the disembodied thong when he saw Pete coming toward him from the end of the ruined hall.

For a brief second, Scott considered stepping back into his room anyway, locking the door behind him. Over the past few weeks, Pete had visited him more than a few times to discuss various complaints that had come his way, usually regarding one or another sorority house that Scott may or may not have upset. Hell, it wasn't his fault if certain girls had unrealistic expectations; if Scott was anything, he was always brutally honest. After all, the walls of his room were painted bright red, and there was a stripper pole suspended from his mirrored ceiling. Even more clear, right behind where his hands were resting on the outside of his door there were Roman numerals imprinted in the wood: XXIII. Although Scott didn't remember anything from the night before, he was pretty sure, based on the thong in his bed, that he was going to have to get out his carving knife and add another number to the door.

Pete reached him just as the Harley skidded its way to the ground floor, aiming toward the main entrance and, beyond that, the front lawn.

"That's something you don't see every day," Pete said, as he peered over what was left of the banister to survey the damage. "Now maybe we can get that damn carpeting replaced. It's soaked up so much spilled beer, it's like walking in a marsh."

"I always chalked that up as part of the house's innate charm."

Pete grinned. "First day I toured this place, they were taking me down a hallway on the second floor when a door hinge came loose and the door swung open—and there was one of the brothers, banging his girlfriend up against his dresser. Whole group of us freshmen standing there in the hallway, and this guy just kept right on going. Yeah, it's hard to imagine anyone making a fuss about the crappy carpet. As long as the girls keep coming through the entrance, nobody gives a damn."

Pete pointed past Scott to the numbers on his door.

"But I don't think I have to convince you."

Scott looked at him with innocent eyes.

"I wouldn't know anything about that. However, you wouldn't happen to have seen number twenty-four wandering around the house somewhere—maybe a little bit naked, save for the odd sequin or two? Unless you're here to break my balls about some sorority house—in which case there wasn't any girl here last night, and she didn't leave her thong in my bed."

Pete shook his head. "Actually, this time I'm here because of your brother."

Scott glanced up, confused. Brent—his stepbrother, actually; Scott's mother had married Brent's father, and Brent had grown up with his stepfamily, saving him from much of the hell that Scott had the misfortune of calling his childhood—was renting a small room on the second floor of the house, on Scott's urging, partly because the house needed the revenue but mainly because Scott hoped Brent would end up at SAE when he matriculated the next year. Scott had always done his best to look after his younger stepbrother; Brent had grown up so dirt poor that the neighbors in his fundamentalist Mormon hometown, just outside of Salt Lake City, would leave baskets of food and clothing on his front porch. Now, Scott's father, Phil, was helping him pay the rent so he could get out of that environment.

Scott himself had only recently reconnected with Phil, who'd enabled him to go to college in the first place, offering him tuition money and convincing him to write a letter to the university describing the harshness of his background, to explain why he might not have had the grades or the opportunities of the other kids. But he'd had no one to look after him when he was a kid, and he didn't want Brent to ever feel as helplessly alone as he once had.

The idea that Brent could be causing any trouble at the frat house—having only just arrived a few days before—seemed laughable. Brent was just about the sweetest, quietest, most humble kid Scott had ever met.

"Don't tell me it was Brent on the Harley?" Scott joked.

"Of course not," Pete said. "But—well, maybe I should just show you."

He gave Scott a minute to grab a shirt and a pair of sweatpants, then led him down the hall to the stairway. They had to walk carefully over the torn-up carpeting and the splintered wood; the tire tracks from the Harley were clearly visible, the rubber burned into an almost cartoonish circle at the top of the stairs.

When they'd made it to the second floor, Pete led Scott to the fourth door down the hall, the room Brent had been assigned in exchange for two hundred dollars a month. As Pete pulled the door open, Scott could see that the room was probably worth only half that; it was little more than a ten-by-ten closet with enough space for a small desk, a twin bed, and a coffee table; a bare light fixture hung from the ceiling.

Brent was at the coffee table, sitting cross-legged on the floor. His appearance was a stark contrast to both his Mormon background and the common SAE fraternity look. Dirty-blond hair rained down over his shoulders, freshly released from a ponytail—strands so grungy-looking they seemed halfway to dreadlocks. He was wearing a torn jean jacket over a hemp shirt, and pants that were so baggy they could have doubled for a skirt. His face was covered in three days' worth of beard growth, and he looked like he hadn't taken a shower in a month.

But Pete wasn't pointing at Brent's appearance. He was pointing at an object on the coffee table in front of him. Scott's eyes went wide. A mushroom—a *gigantic* fucking mushroom, about the size of a loaf of bread—was growing out of the middle of a plastic tray.

Brent looked up and saw them in the doorway.

"Hey, guys. Did you hear something this morning? Like, a motorcycle or something?"

Scott looked at Pete, then back at the mushroom.

"Um, Brent? You doing a little farming?"

Not that any of them had anything against a little weed on a Saturday night, or maybe some shared Ritalin from one of the ADD kids in the house. But a mushroom the size of a small dog?

Brent saw where they were looking and seemed to notice the giant fungus for the first time. Then he laughed, grinning wide. Without a word, he reached out and pulled off a chunk of the mushroom. He held it up in the air, then took a bite.

"It's not what it looks like," he said between chews. "It's for eating. Like, for a salad or something. I mean, it's pretty big, but it's not going to get anyone high."

Scott watched his brother wolfing down the mushroom. Brent was a vegan, after all. Still, he wasn't surprised by the concerned look on Pete's face. Most of the house was like Pete, Garin, and Shane. Good-looking, athletic guys who would probably eventually do well in corporate America. There were some druggies, to be sure, but there was a constant battle between those elements and the sort of country boys who gave the house its heart.

Scott knew that he himself was something a little different. Nobody in the house lived like a Rockefeller—except maybe Shane, whose family had a little money from a tractor dealership—but Scott was pretty sure that nobody else in the house had picked up his mother from rehab nine times by the age of fourteen either. And he was certain that nobody else had ever watched his mother stumble across a trailer, bleeding from deep razor cuts to her wrists. No one else in the house had ever been made to kiss his mother's fresh miscarriage before yet another suicide attempt, this one involving a meat cleaver. Maybe Brent would have to do a little adjusting to fit in, but if Scott could better himself enough to be one of the more popular guys in the house after what he'd been through, anyone could improve himself.

Scott cocked his head toward Brent. "Just try and keep the fungus from eating any of us."

He shut the door and stood next to Pete in the hallway.

"Really?" Pete said. "You think he's going to make it through Hell Week?"

Hell Week was the infamous last seven days before you became a full-fledged member of the fraternity house. Your future brothers spent that time hazing you nonstop to see if you really had the guts and drive to be in SAE. Though it varied from frat house to frat house, there might be things like military-style 4 A.M. wake-ups, with all the brothers gathered around wearing masks, beating drums, holding baseball bats; then you might be dragged out into the snow, stripped down, hosed off, maybe expecting those baseball bats to rain down upon your head. You might be asked to drink cases of beer, then made to do the most disgusting things: clean shit off the floors, march into town wearing nothing but a summer dress, memorize every detail of the house's history and recite it while being pelted with various objects. Your assigned Big Brother acted as a guide and refuge through the process. Scott's Big Brother had been Shane, but even Shane had indirectly gotten in on the action, urging other active members of the house to give Scott a hard time—especially when he noticed how enthusiastically Scott had taken to being hazed.

Then, finally, if you survived Hell Week, you made it into the house. And everything changed from hell to heaven. You were part of the group.

After all that hazing—maybe because of some of that hazing—Scott now considered his housemates the closest people in his life. Shane, Garin, and Pete were really his brothers. He desperately wanted that same experience for Brent. Brent deserved to have people in his life who were loyal and supportive of him, no matter what happened.

"I'll give him the hose myself," Scott said. "And he'll come out smiling."

There was no doubt in Scott's mind that Brent would survive Hell Week and make it into the house. Having a future to look forward to was something Scott and Brent had never experienced before. Given where he and his brother had come from, things could only get better.

CHAPTER 5

LIQUOR UP FRONT—POKER IN THE REAR.

"I guess it doesn't get any more straightforward than that," Garin said as he stood next to Scott, Pete, and Shane beneath the pulsing red-and-white neon sign, surveying the façade of the nondescript building in front of them. "Doesn't look like much from the outside."

Scott grinned, reaching up to put a hand on Garin's shoulder. No, it sure as hell didn't; other than the sign and a couple of beer and Coca-Cola logos above the dark glass windows that ran along the first-story storefront of the two-story building, there was very little to indicate that the place was even open for business—let alone that it was one of the oldest bars in Missoula. The structure was concrete, paneled in slabs that were a mix of peach and vanilla, very 1970s construction. The sign—which read STOCKMANS CAFE BAR and sported a sketch of a longhorn glaring out angrily over the dark downtown sidewalk, daring anyone to step inside—was far from inviting; the motto scrawled in circular script across the darkened plate-glass window—shoulder-high for a regular-size human, waist-high for a giant like Garin—was, on the other hand, pregnant with temptation.

"It's a real game?" Shane asked, sounding a little nervous. "I mean, legal?"

Shane was a pretty straight eagle. Sure, he knew how to enjoy college, was great with the ladies, drank like everyone else—but he rarely got into trouble, kept his room immaculate, and often cleaned up Scott's messes when Pete and Garin weren't around to do the sweeping or explaining. But in this instance he had nothing to worry about. Scott had done his research. He'd discovered Stockman's about a month earlier—nearly a full semester into his junior year and just a few days after they'd welcomed Brent into the fraternity—on a late-night jaunt into Missoula with another frat brother. They'd been looking for cheap beer and even cheaper girls and had stumbled into heaven instead.

"Perfectly legal. It's a single table, nine seats, and they have to keep the pot below three hundred dollars. They've got real dealers and everything."

"What about the players?" Pete said.

Scott was beginning to get impatient, standing out there on the street, people walking by and looking at them like they were fucking tourists.

"A few college kids like us, who sit down with twenty bucks and try to double it by the end of the night. And then a lot of regulars, guys who sit down with a couple hundred and play until morning."

Scott reached for the door before any of the Three Stooges could ask another question. Then he ushered them inside.

The front section of Stockman's was about as inviting as the angry longhorn on the neon sign. The ceiling was low and covered in graffiti, mostly people's names, a few hearts with arrows through them, a handful of mini diatribes against one perceived wrong or another. To the right, a long wooden bar ran down one entire side of the rectangular room, lined with red stools, only about a quarter of which were occupied. To the left, a wood-and-glass wall studded with old photos and framed pictures, everything from horses to cows—if it had hooves, it was on that damn wall.

But Scott wasn't there for the bar or the pictures; he kept moving, quickening his pace lest one of his friends lose his nerve on the way to the rear door, still a good ten yards away. Sure, the place had plenty of cheap beer. Better yet, if you could present a piece of paper that even insinuated that you were twenty-one, you could drink there. But there were plenty of establishments in Missoula where you could drink; this was the only place Scott knew of where you could also play cards.

When he reached the rear door, he ushered his three friends inside.

"Say hello to your new second home."

The back room was only a minor improvement on the front: another low ceiling, more framed cattle on the walls, some actual sawdust on the floor—and an oversize circular poker table situated close to the back wall, surrounded by uncomfortable-looking wooden chairs. The table had a real felt cover, drink holders, and a cabinet built in for the dealer's chips. There was also a slot cut into the table next to where the dealer sat, leading to a box that hung beneath the lip of the felt.

The table was nearly full. All men, most of them in their twenties, a few trending toward middle age. At least half were wearing baseball hats low down over their eyes. None were particularly well dressed; a couple were college kids, like Scott and his friends. A couple more probably worked construction, or were painters, or maybe electricians. Scott doubted any of them were professionals—though in a place like this, nobody was going to ask for your résumé. If you had a few bucks in your pocket, you could play.

A couple bucks—that's all it had cost Scott to get started, the first time he'd wandered into Stockman's that late night about a month ago. The stakes were real low—you could play a hand for as little as a dollar, with the pots topped at three hundred, to stay within the law. That first night, Scott had been more than a little green. The game was Texas Hold'em—the most popular form of poker, the one that nearly every college kid and almost 50 percent of high school boys played regularly, that almost 70 percent of men in the country had played at least once for money, that was practically an American institution, on par with baseball, basketball, and beer. A seven-card game, in which you used five. The play itself was pretty simple. Each player got dealt two cards, then, over the course of the game, five more would be laid out in the center of the table, faceup; each player used three of the center cards to make a hand, and the best hand won. But though the play itself was simple, the *game* was much more complex.

As the cliché went, the game of poker wasn't really about the cards, it was about the players. You had no control over the cards you were dealt—that was pure luck. But what you did with them—or more specifically, how you wielded them against the other players at the table—that was pure skill. Which made the game itself much more about skill than about luck. Over time, a good player would consistently beat a table of bad players, regardless of how the cards were dealt.

That first night, Scott had been anything but good. He'd lost twenty bucks—a small fortune to him, at least one missed meal that week, maybe two—but he'd found himself hooked on the action. The idea that you could look across that table at a total stranger, try to get a read on him just from the way he looked at his cards, or how he fingered his chips, or how he bet, aggressive or cautious or just plain dumb. It was an awesome feeling, an incredible high.

The next day Scott had headed directly to the school library, a place he had seldom gone before. He'd picked up a book called *The Winner's Guide to Texas Hold'em Poker,* and had studied it intensively over the next three days.

Then he returned to Stockman's. Phil, his rediscovered dad, had begun giving him a four-hundred-dollar-a-month allowance—for food, books, and spending money—which had left Scott just enough extra to play cards. And over the course of the next week, he had become a Stockman's regular.

Now he was going to introduce his three best friends to the hobby that was rapidly becoming a welcome addiction. He wished Brent had been willing to come along, but his younger brother hadn't been interested. Since entering the university and joining the house, Brent had been on a nearly 24/7 mission to transform himself, and he was hardly recognizable now. His hair was cut short, and he often wore a tie; he'd even gotten the house involved in a handful of charities, including a soup kitchen in Missoula. Even Pete had admitted that he'd been 180 degrees wrong about Brent.

Scott glanced over at his frat brothers, his three best friends. Shane was still near the door, eyeing the table and stacks of multicolored chips in front of each player with obvious suspicion. Pete was a few steps ahead of him and seemed to be concentrating on the dealer, on the way he counted up the chips in the center of the felt—the pot—and how every few hands he swept a couple of chips out of the pot and flicked them down the slot in the felt. This was the *rake,* the house's little cut, which was a few percentage points of the total bets placed—a few bucks here, a few bucks there. But Garin was already reaching for his wallet. Scott grinned, because he recognized the glint in Garin's eyes. Maybe Pete and Shane could resist the lure of the cards, but Garin was a goner, just like him.

THREE WEEKS LATER
2 A.M.

There was no greater feeling in the world.

It was a frigid night outside, wisps of icy wind drifting up through the floorboards and poorly paneled walls of the crypt-like back-room poker parlor. Pulling a scarf tight around his throat, Scott fought to control his breathing and to contain the spikes of adrenaline that ricocheted up his spine. It was a sudden, primal thrill, hardwired into his nervous system. The minute the dealer had first flipped over the cards, just a few seconds earlier, something inside of Scott had fired off—a chemical reaction surging up from the animal portion of his brain. Anyone who'd ever placed a bet would understand the feeling. Beyond logic, beyond math, beyond statistics and practical thought and strategy was a fire that drove him to play—and win.

It didn't matter that tonight it was mostly his friends—Garin, Pete, Shane, and a couple of his other brothers from SAE—crowded around the felt in the corner of the dimly lit back room. And it didn't matter that Scott was halfway to drunk. He was still lucid enough to recognize the cards—two in his hands, the rest lined up in the center of the table for everyone to see. His cards, on their own, were decidedly unimpressive: a six and a seven, both of clubs. But when you added them to the three cards on the table—that was something else. An eight, a nine, and a ten, also of clubs.

A goddamn straight flush.

That was the opposite of unimpressive. That was something you never, ever saw.

Scott didn't care that the entire pot—a metropolis of colored plastic chips rising like a miniature Technicolor skyline above the center of the round table—came to a little less than sixty dollars. A straight flush was a miracle, whether you were playing for pennies or for millions.

He loosened the scarf a little, letting the cold air bite at his lungs. The other players were watching him, so instead of returning their looks, he watched the dealer—who was in the midst of taking another rake. Just as he had done after every raised hand throughout the night, the man—midtwenties, with a well-trimmed goatee, too many rings on his fingers, and wearing a white-and-red Stockman's sweatshirt with that angry bull stitched across the dead center—swept a few chips out of the pot and slipped them into the slot in the felt. And as usual as of late,

Scott more than noticed the ritualistic move—he found his mind replaying a narrative he'd been writing in his mind for the past week or more.

A few chips to the house, every hand. Pennies against pots made up of dollars—but over the course of an evening like this one, the rake would certainly add up. Scott had struck up a conversation with one of the dealers a few days after he'd first introduced his frat brothers to the bar—and he'd learned that over the course of a year, that one table's rake added up to more than two hundred thousand dollars. The number seemed crazy, impossible. A single table, with a dollar ante, a pot capped at three hundred dollars—two hundred thousand dollars of pure profit a year?

When he'd told Pete about the number—because Pete, a marketing major with a real head for economics, had seemed equally intrigued by the rake—Pete had voiced his own thoughts. If only there was a way they could run a table like that. But of course, without licensing from the state, that would be illegal—a boiler-room operation, illegal gambling, the kind of thing you could go to jail for.

Even so, Scott was more and more possessed by the idea. If one little table in the back room of a bar, limited by the bar's hours and the number of players who stumbled in, could earn six figures a year—what if there was some way to bring the game to more players, maybe many more players, whenever they wanted it—and somehow do it legally? The profit you could make seemed infinite.

At that moment, as the dealer finished with his rake and turned back toward the players, Scott had something more immediate to keep him occupied.

With a flourish, he turned over his cards.

The table went silent. Then Garin whistled low, impressed. Pete stared, stunned. Shane laughed.

Scott grinned, rising halfway out of his seat.

"Pay the man!" he shouted.

With a sudden swipe of the back of his hand, he toppled the metropolis of chips, then scooped them across the felt, toward his own growing pile.

It was a feeling better than sex.

And still, he wondered.

What if you could somehow tap into that on a worldwide scale?

CHAPTER 6

"Look at this, I'm killing this guy."

"I think he's going to cry. Talk about his girlfriend again."

Pete had just made it to the top step of the three-story walk-up, still a good ten feet down the institutional-style hallway from Scott and Brent's new apartment, a few blocks from the frat house, but he could already hear voices, laughter, and the clink of glass bottles. Moving closer, he saw that the apartment door was hanging wide open, allowing the noise to travel unimpeded through the dilapidated building. No doubt the neighbors had grown used to this sort of thing, even though the brothers had moved in only a few weeks earlier. Still, whatever was going on inside the apartment at the moment seemed particularly raucous. He wondered if his friends had gotten into his Ritalin prescription again.

Once he was inside the apartment, he shut the door behind him. There wasn't much to the place—a cubicle of a living room, with a pair of couches that made the ones at SAE look positively regal, a glass coffee table cluttered with beer bottles, various remote controls, and a pair of potted plants that hadn't been watered, leaves fraying and brown, stems dry enough to be smoked. All three of his friends—Scott, Brent, and another fraternity brother named Cal Teller, who often joined Scott at Stockman's on his now regular visits—were gathered around Scott's desktop computer, which had been set up on the plastic bar that served as the dining room table. Scott was at the keyboard, Brent and Cal at either of his shoulders. All three of them were drinking, and from the number of bottles on the coffee table, Pete could guess that they had been at it for a while.

"What the hell are you guys doing?" Pete asked when none of them even acknowledged that he'd entered the room.

"Tell him he's got a small dick," Cal said, still facing the computer screen.

"Yeah," Brent added. "Tell him you're sitting outside his window with binoculars, and that you can see his dick, and that it is very small."

Pete removed his coat, looked for a hook or a hanger, then dropped it onto the floor.

"Really, guys, if that is some sort of interactive porn site, you're all truly pathetic."

Scott glanced back over his shoulder, saw Pete for the first time, and waved him forward.

"No, man, you gotta check this out. This is so cool."

Pete walked across the room and pushed in next to Cal.

The computer screen in front of them was filled with something that was definitely not porn. Whatever the site was, it was extremely rudimentary; everything on it was a dull sandy color, and all the writing was fairly blotchy. In the main area of the screen there was an oval object surrounded by little cartoon-character faces seated in suede chairs, with names beneath each one. And in front of each face, on the oval, a pair of cards, facedown. In the middle of the oval, three cards were laid out faceup.

It took Pete a full minute to realize what he was looking at. It was a Texas Hold'em game in middeal.

Scott was furiously typing—something about somebody's dick and how small it was—and the words were appearing on the bottom of the screen, beneath a squiggly green border. When he stopped typing, another line of text appeared—the response, a lot of angry words, obviously from someone who didn't like being told that his penis was small.

"Is this some sort of a chat room?" Pete asked. "And is that a poker game you're all watching?"

Scott's reflection was grinning in the glass of the computer screen.

"It's a poker game, all right. Real people, playing poker for real money."

Pete laughed. "You've got to be kidding me. You're playing poker over the Internet?"

"Well, no," Scott said. "You have to have a credit card to put real money in. We're just chatting with the douche bags who are playing. There's no one editing the chat feature. It's pretty funny how mad people get."

Pete headed toward the refrigerator to look for beer.

"You guys must be pretty drunk."

"Yeah, maybe, but this is freaking awesome. We've got to do this."

Pete found a Budweiser, went to work on the cap.

"You are doing it. You got a total stranger worried about his manhood. Mission accomplished."

A beer bottle sailed over the top of the computer screen and narrowly missed Pete's head, crashing into the wall of the kitchenette. Scott was out of his seat now.

"No, you idiot. This poker website. It's awesome. I mean, the software really sucks and the graphics are horrible. These are supposed to be palm trees, and the faces all look the same. But this website—it's genius. Even though it's crap, there are like fifty people playing. And they're taking a rake from every table, all night long. They're minting money."

Pete came over to stand behind Scott again. He looked at the computer screen, a little more carefully this time. The graphics really were awful. But now he could make out the palm trees, and he saw what looked to be sand, and beachy waves in a corner.

"What is this site? How did you find it?"

"I made friends with one of the dealers at Stockman's, and the guy told me he'd been making like three hundred dollars a week playing online at this site. It's called Paradise Poker."

"I guess that explains the palm trees."

Scott went back to the keyboard, waxing philosophic once more about one of the players' anatomy.

"I mean," he said, his fingers rattling against the keys. "We would do a hell of a lot better. Come up with something much more sophisticated and clean."

Pete pointed at the cards on the oval table. "You really think people are going to play poker on the computer? More than a handful of dorks with nothing better to do? Poker is a social game. It's about reading the other players, it's about the face-to-face competition. You know that way better than me. You're at Stockman's every week."

"But if I didn't have to leave the house, I could play all the time."

"If you had a credit card," Brent added.

"Well, yeah, obviously. But most college kids have credit cards. And not just college kids play poker. If it were a sport, it would be the most popular in the country. More people play cards than baseball. And worldwide—God only knows how many people play poker around the world."

Pete shook his head; he just wasn't buying it. Most intelligent people he knew would be terrified at the idea of putting their credit card on the Internet. It was 2000; the Internet had a long way to go before most would feel comfortable *shopping* over the computer, let alone playing poker. And then, of course, there was one even more important question.

"Is this even legal?"

"Sure, why not?" Scott said. "Everyone pretty much agrees that poker is a game of skill. And this isn't like owning a casino—nobody is playing cards in your house. I mean, we have to do some research, make sure everything is by the book. I think this website is run out of South America somewhere. But I don't see why anyone would have a problem with a poker website."

Scott went quiet, and Pete could see that he was deep in thought. Maybe he was a little drunk, but he wasn't just kidding around.

Still, Pete was far from convinced. Even if it was completely legal, and you found a way to convince people it was safe to use their credit cards over the Internet, Pete couldn't help repeating what he thought was an insurmountable flaw.

"I just don't see it. Nobody is going to want to play poker over the Internet. No matter how many palm trees you fit on the damn screen."

In his mind, the issue was settled. But Scott's face was still giving off a glow. Pete couldn't tell if it was the result of something internal that had sparked to life, or just the reflection from the screen.

CHAPTER 7

"Ladies and gentlemen, we've begun our initial descent into Seattle-Tacoma International Airport. Please begin powering down any electronics as the flight attendants move through the cabin to prepare for landing."

Scott jerked awake as the cabin lights came on with the last few syllables of the pilot's announcement—nearly upending his tray table with one knee while hastily reaching for the small leather attaché case that had slipped off his lap somewhere over the coast of California. The case felt so damn foreign against his fingers as he lifted it back up; he'd never owned anything that nice before. The material seemed too elegant and sophisticated. On its own, it would've been a great graduation gift, and Scott had been completely shocked when he'd learned that the case was really just an appetizer.

The attaché once again resting securely on his lap, Scott stretched out against the full leather of his business-class seat, careful not to let his boat shoes flip over too far into the aisle to his right. The flight attendants were scurrying about, collecting the last remaining plastic cups from the passengers around him. For the hundredth time since they'd taken off, he couldn't help registering how hot the attendants were—tall, tan, veritable amazons, probably wearing nothing besides bikinis beneath their pale blue uniforms. Scott grinned at the closest of the crew—a staggering blonde who was leaning over the passenger across the aisle to help with a difficult window shade. As she bent, her long skirt rode halfway up the back of her calves, revealing stockings and the heels of a surprisingly sexy pair of red shoes.

A fitting end to an incredible trip. Scott turned away from the aisle, toward the window seat to his left. He was surprised to see that his dad was still fast asleep, and even then, Phil had a wide smile on his face. Though they had reconnected as father and son four years earlier now, it was still kind of amazing to look over and see those features, so similar to his own. Phil was a head taller than him, graying a bit at the temples, but nobody would've had any trouble picking him out as Scott's dad. And probably because they had reconnected as adults, they behaved more like best friends than like father and son.

Which was a big part of why the trip had been so incredible. Under normal circumstances, who the hell would want to go to Rio with his dad? But Phil—that was another story. The attaché case was a nice appetizer, but with Phil along, the trip had been one of the best graduation presents in history. *No-holds-barred Brazil.* Two weeks of pristine beaches, late-night parties, fancy restaurants, and uncountable bottles of fine wine. They took turns playing wingman whenever a string bikini was in sight.

And even with twenty years on him, Phil was almost as good at chatting up girls as Scott. The flight back to Seattle was the longest that either of them had slept uninterrupted in two weeks.

Scott almost felt bad letting his elbow fly over the armrest between them, gently poking at the soft area below his dad's rib cage. There was probably still another fifteen minutes before the plane touched down, but now that the trip was over and they were on their way back to the real world, Scott couldn't wait any longer.

For months now, he had been getting his thoughts in order. He had spent hundreds of hours in the university library. And he was finally ready to take the next step.

"Please tell me the plane had some sort of mechanical problem and had to return to Rio," Phil said, yawning as he rubbed his eyes. "I can't possibly wait until Brent graduates to make that trip again."

Scott laughed, though he knew his dad was serious. Even though Brent wasn't actually related to Phil, the man was generous to a fault. And he had the money to make good on that generosity; he was one of the top investment bankers in the Seattle area. More important to Scott, Phil had built himself up from nothing. He was a true believer in bootstrap ideology—a staunch fiscal conservative who really and truly believed in the American way. Which was why there was no question in Scott's mind about where he had to turn first. With the two of them trapped together in an airplane, even if only for another fifteen minutes, it seemed like the perfect opportunity to make his case.

He carefully unzipped the attaché case and retrieved a stack of computer papers. Then he turned in his seat to face his dad.

"I want to show you something I've been working on," he said, his feet alive against the airplane floor. He could hear the gears in the wings churning; he knew he didn't have much time. But the truth was, if he couldn't make his case in a few minutes, it wasn't going to be something that was worth doing anyway.

"Please don't tell me you want to try and save the world," Phil said.

"We can leave that to Brent and his soup kitchens," Scott responded. "No, this is something a bit more practical. Take a look at this."

He handed his dad the first page from the stack on his lap. It was a screenshot of the Paradise Poker website. Beneath the shot there was a row of numbers, detailing everything that Scott had learned about the site.

As his dad digested what he was seeing, Scott gave him the rundown. His senior year at Montana—just like at colleges everywhere in the country—everyone was talking about Internet ideas. The entrepreneurial spirit had captivated almost every dorm room, and Scott was no different; but he was pretty sure he'd come up with an innovative way to build a company that no one else had yet done right.

"Poker?" Phil said, looking up from the paper. "Scott, I love playing poker as much as the next guy—but as a business?"

"Not just a business," Scott responded. "Big business. International business. Look at Paradise Poker. They've got about a hundred and fifty tables, maybe fifteen hundred regular players. They take a rake out of every game—nearly every minute of every day. Figure just a five percent rake, an average of sixty dollars per pot—that's three dollars per game. If one table deals around one hundred hands per hour, that's three hundred bucks an hour.

Multiply that by a hundred and fifty tables, that's like forty-five thousand dollars a day."

Phil looked at him, then back at the paper. Scott started handing over the rest of his research—calculations based on different gaming parameters, analyses of the handful of casinos that had poker rooms, which were really a tiny minority, because poker as a game of skill wasn't considered a big money earner in Vegas. Even a short history of the game itself: how it had evolved from an eighteenth-century card game played by French royalty, then traveled to the Mississippi steamboats in the 1800s. And he finished with more research into how many people enjoyed the game today: college kids, high school kids, adults, distributed across all incomes and cultures.

Phil took it all in, waited until Scott got quiet before asking the question.

"And you're confident that this is legal?"

Scott nodded. He had spoken to lawyers at the University of Montana and even had a couple of meetings in Seattle. There was a consensus that if there wasn't a law that said you couldn't do it, it was presumed to be perfectly legal. The law that was usually applied to gambling, whether it be online or over the phone, was the infamous Interstate Wire Act of 1961. But every lawyer Scott had talked to was convinced that the Wire Act did not apply to poker. Scott pointed to a page halfway into the stack; on it was printed the entire federal statute. He encouraged his dad to read at least the first few lines:

> *Whoever being engaged in the business of betting or wagering knowingly uses a wire communication facility for the transmission in interstate or foreign commerce of bets*

or wagers or information assisting in the placing of bets or wagers on any sporting event or contest, or for the transmission of a wire communication which entitles the recipient to receive money or credit as a result of bets or wagers, or for information assisting in the placing of bets or wagers, shall be fined under this title or imprisoned . . .

"It's pretty clear that the Wire Act was designed to inhibit games of chance, specifically sports betting. When you look deeper into it—how it was passed and why—this seems even more obvious. Robert Kennedy was trying to take on organized crime, so he convinced his brother to make it illegal to bet on sports. That's what the lawyers say, at least."

Phil leaned back in his seat, clearly impressed. He could see the passion in Scott's eyes, and that was winning him over even more than the research. Phil was a businessman, and he knew that passion was more important than any numbers. You could sell an idea on passion.

"This is some impressive work."

The seat belt light had just gone on over their heads, indicating that they were on the last leg of their approach into Seattle.

"Okay, Scott, you make sure this is legal. And then you do it. I'll write you a check for twenty-five thousand dollars to get you started."

Scott did his best to take the number in stride. To him it was an enormous sum of money. But he knew his dad wasn't just being generous. The twenty-five thousand dollars wasn't a gift; it was an investment. This was Scott's one chance—and he intended to take advantage of it. And to do this right, he wasn't going to be able to go it on his own.

As the plane touched down onto the runway with the screech of tires against pavement, Scott's thoughts were swirling forward. By the time they reached the gate, he knew exactly who he needed to call.

CHAPTER 8

I guess this is what you would call a real upstairs/downstairs kind of operation."

Scott gave Shane a friendly shove through the open elevator doors and followed two steps behind him. Even with the basement's high, vaulted ceilings, the thick carpeting, and the well-constructed cement walls, the sounds of the party upstairs could still be heard, probably drifting down through the elevator shaft. Upstairs, Scott knew, it was all caviar and champagne. A cocktail party in full swing, even though it was barely seven in the evening and his dad had just gotten back from a business trip overseas. But down in the basement, it was a different scene altogether.

The basement was large and rectangular, with no windows, no paintings on the walls, and almost no furniture. Shane had stopped a yard from the elevator, which had since closed behind them; it seemed like Scott's frat brother was fighting the urge to turn and try to run back upstairs. The basement certainly didn't look like the office of a fledgling Internet company—maybe more like some sort of terrorist hideout. There was a whiteboard in one corner, covered in fairly arcane computations and sketches, and stacks of papers like the teetering walls of a hastily constructed fort, set in a semicircular pattern around a pair of matching computer stations. Both monitors were on, dueling screens of sand and green. Without a doubt, Shane immediately recognized the squiggly palm trees from Paradise's website, even from that distance.

"I really like what you've done with the place," Shane said. He pointed toward a pair of closet doors along the back wall of the room. "Is that where we keep the hostages?"

"Actually," Scott said, crossing to the closest stack of papers, "that's where we sleep. There's a cot in each closet."

"You've got to be kidding me."

Scott grinned. He wasn't kidding in the least. And he knew that despite Shane's rumbles of discontent, his friend was completely on board. The basement was a stark contrast to the incredible mansion upstairs—but it was all theirs. When Scott's father had offered him the use of the space for his fledgling company, he'd jumped at the chance. The beautiful home above it—a sprawling estate on Lake Washington with more bedrooms than Scott could count, manicured grounds, and even a fully operational golf course where his dad held one of the area's premier annual charity events—acted as aspirational motivation. Scott intended to own a house like that one day. This basement operation was how he was going to get there. Besides, it wasn't going to hurt that his dad would be entertaining wealthy colleagues and clients upstairs. Scott had agreed not to pitch himself to

Phil's clients in any overt way—but if someone wondered what the heck was going on in the basement, well, Scott would be more than happy to give him a tour.

Even with living out of the basement, twenty-five thousand dollars wasn't going to last very long. If Scott was going to make his company a reality, he was going to need to raise more money. And before he could go after additional funds, he needed personnel.

Shane had been an easy first choice. Scott trusted him implicitly, knew he was a hard worker—it didn't hurt that he was the most anal-retentive of the fraternity bunch—and his eye for detail would be a great help. Equally important, his family had a tractor dealership; they had the ability to invest in a new business venture. Most important of all, Shane had coincidentally already moved to Seattle for a job that hadn't worked out—which made him eager and willing. When Scott had pitched him the idea over lunch at a kitschy place called Chang's Mongolian Grill, Shane had been enthusiastic from the beginning; when he'd opened his fortune cookie at the end of the meal and read his fortune out loud, he'd been completely hooked: "A confidential tip will clue you in to a great financial deal." Scott couldn't have planned it better if he'd tried.

Scott pulled one of the sheets of paper off of the closest tower and handed it to his friend. Shane saw that it was a list of names—all of which he recognized.

"And here I thought I was just really, really special. It looks like half of our fraternity is on this list."

"Just the best, brightest, and those who come from a little bit of money. Look, it seems pretty obvious to me: Why go knocking on doors when we have this incredibly deep bench—SAE?"

"Have to admit, down here in the basement it does feel a little bit like we're back in the fraternity house."

There was even a pile of beer cans in a corner, along with a row of empty Red Bulls. Why not extend the frat house feel? The SAE house was one of the greatest experiences of both their lives. There was no reason to change a model that had already been proven to work.

"Who are you going after first?" Shane asked. "Pete? Garin?"

"I already spoke to Pete. He still isn't on board with the idea of people playing poker online. And also, well—he did just get married."

At least Pete had agreed to act as an informal consultant as they moved forward, but he wouldn't be moving into the basement. He'd married his college sweetheart, and she probably wouldn't have liked living in a closet anyway.

"And Garin?"

Scott grinned. "He's driving his Mustang up next week."

"And who else? You're not going to get Brent to drop out of school early, are you? He just got elected president of the house for next year."

Scott shook his head. He was damn proud of his brother—it was amazing, the transformation the kid had undergone. Now he was going to be president of the whole goddamn frat. Scott watched as Shane lowered himself in front of one of the computers and started poking around the keyboard. Now that he had a little money, Scott had opened an account on the website—so they could play along and see how flawed and imperfect Paradise Poker seemed to be. He was already coming up with ways the site could be improved. But to get there, they needed money—because it was money that would lead to new software.

None of the SAE brothers were computer programmers. In fact, none of them had ever taken any computer courses at all. Which probably made them the least qualified people to start an Internet company.

But that would soon change. Scott was determined to put together the perfect team. They wouldn't leave the basement until they had a working business plan and an avenue to the software that would make it all sing.

Garin was the next piece in that puzzle, and even though he had only been hired a day ago, he had already proven himself invaluable with a simple suggestion of who they could go after next.

Scott was going to let Garin make the call himself.

That's pretty cool," Garin was saying as he squatted against a beanbag chair in the corner of the basement, a cordless phone held in the crook of his neck. "A whole semester abroad in Paris? Can't imagine what that would be like. I'm not even sure I could find the place on a map."

Garin looked up from the phone and across the room at Scott and Shane, who were seated next to each other at the computer stations. He gave them a thumbs-up. A semester abroad in Paris was the kind of thing that none of them could ever have contemplated; it was obvious that the kid on the other end of the line would make an important addition to their team.

"Welcome back," Garin continued. "I know it's been a while since you left Montana, Hilt, but I've got a little proposition for you. Shane and Scott are here too, and we're working on a poker business. Yeah, online. It's an Internet company, where people can play poker."

Garin launched into the pitch that he, Scott, and Shane had developed—a sort of mini business proposal that they had put together from the research they had compiled. Scott could tell, even from across the room, that Hilt was at least listening. If he liked what he heard, they'd be in great shape.

Oscar Hilt Tatum IV had attended the University of Montana for only a single semester—racing back to Florida, where he had grown up, because he hated the cold—but in that short time he'd made quite an impression. He'd rushed the SAE house, gotten accepted in no time—and then had shown up with a BMW M3 convertible on the back of a truck. That sight still stood out in Scott's mind years later.

Hilt came from money. His parents were prominent in the St. Petersburg medical community, and his family extended deep into the professional field. Hell, he had a Roman numeral after his name. The only Roman numeral Scott had ever been involved with had been carved into his frat-room door.

If any of them aside from Shane had access to people with money, it was Hilt. When Garin finally finished his conversation and hung up the phone, Scott could hardly stay in his seat.

"Did he seem into it?"

Before Garin could even answer, the phone, still in his hand, started ringing. Garin stared at it, then finally put it to his ear.

A few seconds later he hung up, then rose to his full height and clapped his palms together.

"He said he'll be here in three days. And he's going to bring a hundred thousand dollars in investment money with him. He wants in, and he wants a piece of the company. I think we damn well better give it to him."

Scott tore across the room and caught his friend in a grip that was half tackle, half bear hug.

They had just quintupled the value of their company with a single phone call.

The next six weeks flashed by at ten thousand RPMs, bolstered by a constant stream of Red Bull, adrenaline, and a shared determination to one day get the hell out of that basement and onto a bigger stage.

Very quickly there emerged a strict daily routine. Scott, Shane, Garin, and Hilt were all at their computer stations every day by 7 A.M. Punching keys, doing whatever research they could, writing away at the business proposal that was growing line by line, paragraph by paragraph—all through the day, until 9 P.M., when one of them would break first, sliding off a stool or beanbag chair and onto the floor, ready to crawl toward one of the cots in those damn closets. None of them had ever worked so hard. Entire weeks went by without any of them stepping outside. It got so bad that eventually they decided to create a ritual night out: Wednesday, because Wednesday was ladies' night at a favorite bar in downtown Seattle, a place with *Billiards* in the name.

As a team, they had learned to function even more efficiently than Scott would have thought possible. Once Hilt had arrived in the basement, he had immediately been assigned the task of headlining their money-raising efforts. Polite, soft-spoken, and slight of build, especially compared with Garin, he had an amazing affinity for numbers and all things economic. He was also an intensely logical person who spoke faster the more excited he got. He quickly developed an incredibly convincing pitch of his own, which he plied over the phone as often as possible, starting with family friends down in Florida and extending through the fraternity network to anyone he thought might be willing to invest.

They made their goal simple: $750,000, which they intended to raise within three months. That, they believed, was the minimum amount they would need to launch their company. Garin, Shane, and Scott, who were focused on editing and constantly revising their business plan, had come up with the number by both analyzing Paradise Poker's financials and extrapolating using what data they could find about the market as a whole. In the year since they'd discovered Paradise Poker, more companies had entered the business, and a couple in particular were growing at a fairly rapid rate. PokerStars, run out of the Isle of Man, was well capitalized and seemed like it was going to rise to the top of the heap. Another, Party Poker, was growing by leaps and bounds, and was also well financed—its founders had made a pile of money on 1-900 sex lines and had poured that capital into a first-rate poker site.

The one common denominator that they had found among the sites was that they were all based outside of the United States. Even though every lawyer they met with continued to assure them that there was nothing inherently illegal about running an online poker website, it seemed that all of the companies were being run overseas, even though the large majority of their customers were American.

Scott and his team hadn't yet come to any conclusions, but

they began to see the many benefits of launching their company overseas: cheap labor, governments that were okay with licensing a gaming website, experienced platforms. If they were going to run an international business, there was no reason not to think internationally.

Before any of them would be getting on a plane, however, there was one more pressing issue: they needed to come up with a name for their website. On the Internet, your future was only as strong as your domain name. It was more than just words on a monitor; it was your location, your home, and your brand. Paradise Poker, PokerStars, Party Poker—they were all strong choices, because all of them left you with a feeling, an emotion. Paradise—that was self-explanatory. PokerStars gave you a feeling that just by playing there, you were some sort of poker celebrity. Party Poker—well, wasn't that what it was really all about? An online party with friends and strangers that never had to end.

Scott wanted something just as powerful. Something sophisticated, something that brought to mind a classy operation, a place where you might have a martini and a cigar and play a round of poker.

But despite their efforts, a good name eluded them. In recent days they had grown so desperate, they had taken to leafing through the dictionary, just throwing out words, adding *poker.com* to whatever they found. CallPoker.com, JackPoker.com, PlayPoker.com. Nothing seemed good enough. Before they did anything else, they had to solve this problem.

An Internet company without a name was like a bar that no one would ever be able to find.

It was a Wednesday night, a little after ten o'clock, and the BMW 5 Series sedan was positively throbbing. Techno music reverberated through the speakers built into the dashboard, making the very windows rattle as Scott navigated the sleek automobile down a dark stretch of highway. Trees were flashing by on either side and there were mountains in the distance, but it was hard to concentrate on anything other than the techno. Garin and Hilt, in the back, and Shane, in the passenger seat, had all been complaining about the music since they'd pulled out of Scott's father's driveway. Scott had left it on just to spite them.

It was Garin who'd made the obscene suggestion that the radio had been left on a techno setting by one of the girls Scott had brought home the week before. Scott was seriously offended by the idea that he would sleep with a girl who liked techno; he was pretty sure that it wasn't one of his conquests to blame, because his dad had been on quite a tear recently. The blonde whom Phil had brought home a few days earlier was wearing the kind of high heels that would have fit in well at a rave.

So he left the music on, to punish Garin for his comment, and was fully enjoying the looks of pure agony on his friends' faces in the rearview mirror. They had been driving for thirty minutes, which meant there was still a good ten minutes to go before they reached the billiards bar and ladies' night. Unless one of them picked up a girl with better taste in music, Scott was going to make sure the techno continued for the ride home.

Five minutes later, he had grown so used to the bitching of his passengers he almost didn't hear that one of them was shouting at him from the backseat. It wasn't until Hilt repeated what he was saying a third time that Scott realized he wasn't yelling about the music. Scott immediately reached for the volume, and the car went dead silent.

"Absolute Poker!" Hilt shouted again.

The words reverberated off the windows and leather seats of Phil's car. Eventually, Shane spoke.

"It's not bad."

"Sophisticated," Garin added. "Cosmopolitan, kind of a lounge feel. You think there's any way something that simple is still available?"

Scott could feel the engine of the BMW pulsing through the steering wheel beneath his fingers.

"Only one way to find out."

Without another word, he yanked hard on the wheel, sending the car careening into a skid. He made the U-turn by inches, the two right tires spitting up gravel, grass, and pavement. Ladies' night was instantly forgotten.

Twenty-five minutes later the four of them were hunched over one of the computer stations as Scott punched in the words. It took less than five seconds before they got a response.

Scott leaned back and lifted his hands into the air. His friends high-fived behind him, and then he quickly punched in the information to buy the domain name. Twenty-nine dollars to lock it down—and AbsolutePoker.com was officially born.

CHAPTER 9

"That's got to be him."

Scott shielded his eyes from the late-afternoon sun, which seemed to hang like a vast and flaming Christmas ornament just inches beyond the mostly open-air glass entrance to the airport. Trying to see where Shane was pointing, he could make out the mass of taxi drivers, swarming like flies whenever anyone who looked even remotely North American exited through the revolving doors. Scott himself had briefly stepped out to the curb before being beaten back by the aggressive cabbies; for now he was satisfied to stay in the safety of the baggage claim area with Shane, Garin, and his dad. The trip to Brazil—his first outside the United States—had certainly opened his eyes to how different a foreign culture could be. But even from his brief moment outside, his first tentative steps into the Central American country of Costa Rica—lost in the jumble of drivers shouting at him in Spanish and broken English, the thick, humid air catching in his throat and filling up his lungs, the smoggy scent of the nearby urban jungle that was San José, the country's capital—gave him the feeling that this place was a world apart.

"Most definitely," Garin said, in tune with Shane. "There can't possibly be two people who look like that in this hemisphere."

Then Scott saw him too, pushing his way through the crowd of taxi drivers and into the revolving door—and whistled low. Eric Tuttle was truly something to behold. Elongated to an almost comical extreme, with gangly limbs like a humanoid spider and spiky red hair above a paper-white forehead, he snaked forward in a gray business suit with wide, anachronistic lapels and an even wider eighties-style tie.

Eric saw them pointing in his direction as he broke free of the door and entered the airport lobby. He smiled, revealing a set of oversize veneers, and raced toward them at full speed.

"Now, where the hell did you find this guy again?" Scott's dad coughed as he leaned back against his designer suitcase. He was the only one of them who had packed a case—the rest had small duffel bags slung over their shoulders. To Scott, seventy-two hours in a tropical country meant three T-shirts, three pairs of boxers, one pair of jeans, and a box of condoms. But his dad was decidedly more urbane.

"The Internet," Garin said, grinning. "Of course. Lots of websites talk about him—he's supposedly the best gaming consultant in the area. Supposedly works with most of the sports books, and people say he used to be a little involved with Paradise when they first opened up here. He's also pretty cheap."

"You had me at cheap," Scott interrupted. "Now, shut the fuck up. We've got three days to learn everything we can about this place, and he's as good a guide as any."

The recon trip had been Scott's idea; even though they were still far from reaching their financial goal of $750,000 in investment seed money, they were at the point where they needed to make some firm decisions. First on that list, now that they had a company domain name, was settling on a location for their headquarters—a home, as it were, where they could incorporate and begin building their brand.

Costa Rica seemed the natural first choice. Paradise Poker was located there—and in addition, the country seemed to be ground zero for the online sports book business. Which meant there would be a lot of experienced people who knew the tech and the industry. As with many Central and South American countries, Costa Rica could also provide lots of cheap labor—but in Costa Rica, that labor would be well educated. From the research they had done, Scott had learned that it would be fairly easy to get incorporated in the country and to secure a gaming license. The location lent an air of credibility. And from the pictures he had seen in the guidebooks Garin had brought home from the Seattle library—well, it didn't hurt that the place was a tropical paradise. The idea that they could all move there—start their company, breathe life into AbsolutePoker.com in such an exotic locale—it was pretty fucking exciting, and very fucking cool.

The bizarre-looking consultant was breathing hard when he finally reached them.

"Welcome to Costa Rica, gentlemen. We have a full schedule, so let's get moving. The car is just outside. Ignore the taxi drivers, they're just part of the place's native charm."

And with that, the man spun on his heel and strutted back in the direction from which he'd come. Scott glanced over at his team; they all seemed about to crack up, and he scolded them with his eyes. This was serious business, and they were supposed to be acting professional. Even if, from the looks of their skeletal consultant as he moved away, they were about to embark on something akin to Mr. Toad's Wild Ride.

Eight hours later, the ride was beginning to feel a lot more like a merry-go-round than a roller coaster. The five of them were jammed into Eric's bright orange Fiat, speeding along palm-tree-lined highways, with Eric all the while aiming a hand left, then right, then forward, pointing out buildings that ranged from low, boxy warehouses, to ranch-style houses, even to the odd multistory apartment building. Each one, according to Eric, was the home of a sports book or an online casino. They asked him repeatedly about Paradise Poker, and eventually, as they were riding along one of the steeper roads leading up to the base of the hills above San José, he pointed toward a two-story house with gated windows and white shutters, mostly hidden behind a high security fence. He told them that the guys who ran Paradise Poker were essentially shadows in San José; everyone told stories about them, the gringos who lived up in the hills and rode around in black Escalades, throwing money around like it was toilet paper, always traveling in packs protected by bodyguards, surrounded by girls. Nobody ever really saw them, and who knew if the stories were even true? But supposedly, these were their offices, behind that security fence.

Unfortunately, Paradise wasn't one of their destinations that day. Instead, Eric had arranged for them to meet with a slew of sports book owners who were bringing in the lion's share of business in the gaming industry. And even though Scott had repeatedly explained that they weren't interested in sports gambling, Eric had maintained that the sports books were the place to start.

So again and again, Eric parked the Fiat in front of one of the nondescript warehouses or the low ranch houses and ushered the four of them inside. Each time the setup was the same. Cubicled call centers spread out across bland spaces—hell, if you walked into a call center at Hewlett-Packard or Cisco, you'd expect to see the same thing. Once they got into the back offices, they found that most of the operators behind the sports books were Americans, while the front-office staff was usually Costa Rican.

But the most remarkable thing about the sports books—and the thing that they all seemed to have in common—was the seedy element at the top levels. Most of the American operators seemed like criminals—the way they dressed, the way they spoke, the way they offhandedly mentioned associates back in New York and Vegas. By the third and fourth book they'd visited, the seediness was reaching almost cartoonish proportions.

Around 4 P.M., at the last stop before they were to break for dinner—and start a night of festivities that Eric had assured them would rival anything they had experienced at the fraternity house—they pulled up in front of a warehouse at the edge of an urban sprawl of similar rectangular buildings. Eric parked the car and led them through the front door, past a security desk, then a pair of secretaries who didn't even bother looking up from the Spanish newspapers they were reading. Then through an unmarked wooden door and into a corner office. And there, the man behind the desk was right out of a Martin Scorsese movie.

Overweight, in his midfifties, with an angry, pug-like face

and a ring of graying curls barely covering the expansive dome of his skull, he was wearing a polyester suit right out of the seventies, all brown and burnt orange, and he was holding the biggest cigar Scott had ever seen.

The minute they sat down, the guy started in on them—the same broken-record song and dance they'd heard from every sports book owner they met. *Poker is a lost cause. There's no money in poker. Sports gambling is where it's at. You're going to lose every penny of your investment money . . .*

When Scott pointed out that running a sports book as an American citizen was clearly illegal, that sports betting was clearly against the Wire Act, the man just brushed his concerns aside.

"It's a new era. The Internet changes everything. We're not bookies, taking bets off of some pay phone in the back of a bar. This is the Wild West. And I don't see any sheriff knocking at our door."

As he spoke, Costa Rican employees filed in and out of the room, putting papers in front of the man for him to sign. Most of the time he just waved his cigar at them, only pausing now and then to add his scrawl to a paper he deemed important enough to warrant his attention. Scott had no idea who the employees were, but he got the feeling from the way they were dressed that at least a couple of them were lawyers.

His suspicions seemed to be confirmed when Garin pulled a small tape recorder out of his pocket and placed it on the man's desk in front of him. Garin had been using the tape recorder to help them keep track of everything they were supposedly learning, but it was the first time he'd taken the thing out midconversation.

Almost immediately, all hell broke loose. The two Costa

Ricans who happened to be in the room at the time began shouting in Spanish, one of them waving his papers so violently it seemed he might take flight. The sports book owner looked up, saw the recorder, and jabbed at Garin with his cigar.

"You want to holster that, buddy? My guys get a little antsy around wires."

The type of people who referred to a tape recorder as a wire weren't usually paragons of good business practices—and for Scott that pretty much summed up the visit. The sports book business was shady, and it didn't look like it had changed much since Robert Kennedy had gone after it with the Wire Act.

Scott wasn't interested in sports betting. Poker was his interest, his passion—and that was all Absolute Poker was going to do. If there wasn't real money in online poker yet, it was simply because nobody had done it right.

"I don't think we're in Montana anymore."

Scott would've traded every colón in his pocket—and half the blackjack chips stacked in front of him at the semicircular gaming table—for a photo of the expression on Garin's face. Garin had gone from young American businessman in a shirt and tie to shocked farm boy in the space of less than two seconds. Scott couldn't blame him; it wasn't so much that the girl had reached out and grabbed Garin's crotch—it was the nonchalant way she had done it, as if it were the most normal thing in the world. She had been walking by their table, hanging on the arm of a guy who looked like he was at least sixty, and she had just reached out with a smile and given Garin a little squeeze.

Even more bizarre—the guy on her arm hadn't cared. In fact, he'd laughed, and given Garin a thumbs-up.

"You weren't kidding about this place," Phil said. Scott's dad was seated to Scott's left at the blackjack table, Eric standing right behind him, hands crossed against his narrow stomach. The dealer was in the midst of shuffling, but even he cracked a smile.

"There's something for everyone at the Del Rey," Eric said, quite seriously. "People come from all over the world to take part in what's on offer here. Which, if you haven't guessed, is just about anything."

Scott had to admit, Eric hadn't been exaggerating when he'd promised to show them a time they'd never seen before. When they'd first pulled up in front of the seven-story, 1940s-style pink building at the corner of a narrow, crowded street in San José, Scott hadn't expected much. But once he and his crew had pushed their way through the street urchins, past more of those damn ever-present taxi drivers, and finally through the glass entrance into the Del Rey's lobby, he could see that Costa Rica was going to leave Rio in the dust.

With three bars, a casino, a dance club, and a restaurant across the street, as well as a 108-room hotel above, the Del Rey might have looked like any retro Central American resort in a brochure, but three steps into the place, Scott could already tell that it was much more than that. The front area had a sort of tropical sports-bar feel, with soft couches, carved mahogany furniture, potted plants, and televisions blaring from every wall. But the clientele was mostly women—and damn, every one of them was eyeing Scott and his group with palpable intensity. Hot pants, tight jeans, belly-baring halter tops, hair spray, bright red lipstick, and so much silicone you could take your eye out if you weren't careful—there was no question in Scott's mind what this place was all about.

Ahead of the lobby area was the cashier's cage, and next to that, a floor-to-ceiling mirror—in front of which stood a few more girls, checking themselves out, making minor adjustments for the night ahead. Beyond that, the small casino, filled with table games, a roulette wheel, and a handful of slots. On the other side, the Blue Marlin Bar, which, Eric reported, had the hottest bartenders in all of Central America. And beyond that, the hotel reception desk, staffed by a handful of Costa Rican natives. Eric explained, as they pushed forward, that it was ten dollars to the hotel for each girl you took upstairs—and around a hundred bucks more to the girl, though that was often negotiable.

"On a good night," Eric said, "there could be two hundred girls in here, all for the choosing. From Colombia, Panama, Dominica, even Eastern Europe."

"Holy crap," Shane responded, taking a deep swig from his beer bottle. He had been drinking since they'd sat down at the blackjack table; they'd chosen the relative calm of the casino portion of the resort, because it seemed that the girls were mostly congregated in the front lobby, the bars, and by the hotel desk. But obviously, with so much talent in the place, nowhere was really off-limits.

"It's like Disneyland for whores."

"In that analogy," Scott said, "I think we're the whores."

Shane reached out and grabbed another passing girl—about five foot five, built like a water slide, with blond highlights, a denim skirt, and glitter shining from every inch of exposed skin.

"Hell, yes, we are!"

Scott had to laugh. It was amusing seeing Shane—usually the most self-contained of the group—losing his shit like that. But it wasn't surprising. This place was the new Wild West, and it seemed like anything was fair game. The girls, the gambling, the booze—as Scott focused more closely on his surroundings, the more he let his eyes adjust to the frenetic motion, the deeper he could see into the crags and corners of the place. Girls handing off little paper bags in exchange for a handful of bills, customers palming plastic-wrapped cubes that were either green and leafy or white and powdery—even the odd plastic pipe, shoved into a back pocket. From the guidebooks Garin had shown him, Scott knew that unlike prostitution, drugs were illegal in Costa Rica. But from what he could see, just sitting there at a blackjack table in the most festive bar he could imagine, the place looked pretty damn lawless.

He glanced over at Shane, half off his seat, the girl he'd grabbed now draped across his lap with one hand gripping his thigh. He looked at Garin, who was chatting up a pair of Colombian girls who could have been sisters, sporting matching red hot pants, leather boots, and strikingly identical silicone bolt-ons. He saw his father, leaning away from the table as he lit up a Cuban cigar.

Christ. Building a business here was going to be a unique experience, to say the least. Scott and his friends unleashed in a place with no restrictions, no rules—it was more than a little terrifying to think about. But it was also kind of perfect.

He wasn't opening a hardware store; he was launching an online poker site. He was there to break ground. He intended to turn an industry on its head, build a business around a game that everyone was telling him couldn't make money. The energy around him, the Wild West feel—it was exactly what he needed.

Poker, the way it was played in America, had been born in Mississippi in the Wild West era. But Scott intended to take it into the modern age, to turn it sophisticated, to make it as tempting and addictive packaged in electronic bits and bytes as it was on a felt table over a sawdust floor.

Sports betting—that was a different business. It was call centers taking phone calls all day long; it was dirty, mobbed up, and illegal. Poker was sophisticated, young, and hip. To capture that, Scott knew the key was going to be the game itself—the software.

Now that he had his core team, was on his way to finding financing, and knew where he was going to build his empire, the next step was to figure out how.

CHAPTER 10

"Mission accomplished," Garin said as he rejoined Scott in the waiting area of Sea-Tac's international terminal. He pulled a plastic bag out of his duffel, which was slung over his left shoulder because his right shoulder was still bruised up from an impromptu pickup basketball game they'd gotten into in a corner of the basement the night before while Hilt had booked their last-minute plane tickets—*coach, damn the cheap little bastard.*

"Blue Label, baby. Class all the way."

Garin lifted the neck of the bottle of Johnnie Walker out of the bag so that Scott could nod his approval. Two hundred bucks, but if the information Garin had pulled off the Internet was correct, it would be money well spent.

"If we can keep our hands off of it for the fourteen-hour flight, at least we won't walk in empty-handed."

Garin reached into the bag and pulled out a second item—a Seattle Mariners baseball cap, emblazoned with the number 51.

"And this too. An Ichiro Suzuki hat. They're gonna love it. Hometown hero and all."

Scott stared at his friend.

"Dude, Ichiro is Japanese. We're on our way to Korea. They're different fucking countries. And I think they hate each other."

"Shit." Garin tossed the hat back into the bag. "We'll leave it on the plane."

"You can be pretty stupid sometimes."

"It's because I'm so athletic," Garin joked. "I never had to do no learnin'. Seriously, Korea, Japan—aren't they all Asian?"

The airport intercom coughed to life above their heads, letting them know that their flight—Continental, nonstop to Seoul, Korea—was getting ready to board. Scott reached for his own duffel, tucked between his feet against the dull green carpet. It felt light—even for a forty-eight-hour trip, most of which was going to be spent in the air. One professional outfit, including a single dressy shirt—that was pretty much it. Like the recon mission to Costa Rica, this wasn't a pleasure trip; it was pure business. And besides, they were on a shoestring now; as much as he wanted to give Hilt a hard time for the coach tickets, he knew they couldn't blow any of their budget on extraneous expenses. Especially after the check they'd just written.

Fifty thousand dollars. Even now, days later, after Scott had gotten the chance to digest the number, it still seemed insane. A fifty-thousand-dollar check, made out to some dude in Korea they hadn't yet met, who ran a company they knew very little about. But the little they did know had forced them to move forward—because without the Koreans, they had nothing but a domain name.

It was Shane who had first made the connection with the Koreans; he'd been searching the Internet for poker and gaming software and had kept coming up with addresses in Seoul—over and over again. It seemed that most of the good software was being written in Korea. Eventually he'd narrowed down his search to a company run by two brothers—C. J. and Christian Lee. The materials on their website looked pretty good, and on the phone, C.J.'s English was almost flawless, and his presentation good enough to convince them that he was competent, somewhat experienced in gaming software, and, most important, eager to help them launch a unique site. The thing was, he had insisted on being paid up front—fifty thousand to start, with another fifty thousand when he delivered what they were looking for—the beta software—and then yet another fifty thousand when the software was complete.

A small fortune, but Scott knew that the software was going to make or break them. So he'd written the check. It was half of the money Hilt had brought with him to the company, but Hilt had assured his partners that money begot money. Already, their compact business guru had made inroads to a handful of other family members and friends down in Florida, who had reacted positively to their work-in-progress business plan. And even more significant, Shane's mother and uncle had made critical investments, totaling a quarter of a million dollars. They now all felt sure that with a good software package to present to potential investors, they would quickly be able to reach their goal of $750,000.

All of which meant that Scott was going to spend fourteen hours in a tiny airplane seat fighting the urge to down that bottle of Blue Label. According to Shane and what he'd found on the Internet, it was a tradition in Korea to give a gift at a business meeting, and supposedly Koreans were nuts about scotch. They probably weren't nearly as nuts about Japanese baseball players. Still, even though he could be a fool sometimes, Garin was the best choice to go along for the software meet and greet. Hilt was the business guy, Shane the conservative research man. Scott was the ringleader, the showman, the guy in the top hat. And Garin was the meat and potatoes. Sometimes more potato than meat, but at least he was always enthusiastic.

"Just for the hat," Scott said as he led Garin toward the line of people that was already forming at the entrance to their gate, "I'm taking the window."

"Go right ahead. I'm gonna try and flirt my way into first class. Maybe one of the stewardesses is a big Ichiro fan."

"Just don't get yourself arrested before we reach Seoul," Scott said with a grunt. "We've got about fifty bucks in traveler's checks to last us through two days, so nobody's getting bailed out of airport jail until we get our damn software."

Three A.M., Korea time, and they hit the ground running. Thankfully, there was a uniformed driver with a sign waiting for them after they passed through Customs, because otherwise they'd have been completely lost. Every sign they passed was in Korean—squiggles that might as well have been hieroglyphics, because even with a Lonely Planet guidebook, neither one of them could have translated their way to a subway.

The airport was a good hour from the city; the modern highways that looped through tunnels, down ramps, and along overpasses might as well have been northern New Jersey, for all they could see out the sleek sedan's tinted windows. And then *blam*, there it was: a modern pincushion of frighteningly bright lights, towering skyscrapers, glowing bridges, crowded boulevards—and so much goddamn neon the whole place looked like a spaceship that had crash-landed, flipped over twice, and caught on fire. It was beautiful, strange, foreign—and totally futuristic. Scott was jet-lagged, his head still throbbing from the canned, dry airplane air, his legs cramped from the coach seat—but looking at the lights of Seoul, he felt his insides come alive.

A kid who'd grown up in a trailer at the mercy of a mentally disturbed, majorly addicted single mother, literally dodging frying pans, irons, razor blades, and the odd shotgun blast—and here he was, speeding through the streets of Seoul, in the back of a Mercedes sedan, on his way to a business meeting. All those hours spent jammed in that basement downing Red Bull and crunching numbers were suddenly worth it; this was really happening.

He looked at Garin, whose face was striped with reflected neon.

"This is *Blade Runner* shit," he murmured, and then he pressed his face against the side window, letting the rumble of the Mercedes's engine play deep into his bones.

Floor-to-ceiling windows; high-tech, chrome-and-leather ergonomic office chairs; thick, blindingly white wall-to-wall carpeting; giant glowing TV screens hanging from the ceiling; and a burnished redwood boardroom table running the entire length of the room, supporting a half dozen state-of-the-art computer monitors and keyboards.

"This is quite an operation," Scott said as he followed two steps behind C. J. Lee's mechanized wheelchair, flanked on his left by Garin, whose jaw was down to his chest, gawking at every damn detail like he'd just stepped off a tractor, and on his right by Christian, C.J.'s younger brother. "And everyone in here is going to work on our account?"

There had to be twenty people in the place—at least eight in the boardroom, already waiting for them, gathered around the monitors, some clattering away at the keyboards, and another dozen or so scattered throughout the rest of the fourteenth floor of the glass-and-chrome skyscraper in downtown Seoul—where the Mercedes had taken them after they'd showered, changed, and gotten a couple hours of downtime at the Hyatt, where C.J. had insisted they stay.

"Software is a lot more labor-intensive than people realize," C.J. responded. His English was near perfect, which made sense, because, as he'd explained over coffee in his corner office, down the hall from the boardroom, he'd gone to high school in L.A. and had even spent a few years at Berkeley developing an obsession with software development. "And gaming software has its own unique complications. But we're very good at what we do."

Christian nodded vigorously, a cigarette dangling precipitously from the corner of his thin lips. Nearly everyone they'd seen was smoking—and not just socially, not just a cigarette here and there. These guys were chain-smokers, constantly pulling packs out of back pockets, tossing butts toward the numerous ashtrays situated on every windowsill, desktop, and computer tray. The air was thick with smoke, but somehow it didn't bother Scott—maybe because of those windows, and the spectacular view of downtown Seoul, which made the place feel airy and clean, even though he was basically breathing in pure exhaust.

"Complications, complications, complications," C.J. continued, navigating his wheelchair to the head of the long boardroom table. He waited for Scott and Garin to sidle up next to him, then nodded at Christian, who grunted and said something in Korean to the other employees. The door shut behind them, and all of the employees rose from the keyboards and monitors, standing at a sort of attention, hands clasped behind their backs.

Scott felt like he was back in college, surrounded by new fraternity pledges. He watched as C.J. touched buttons on the armrest of his wheelchair, moving himself close enough that he could just barely reach out with a finger and touch one of the nearby keyboards.

Scott and Garin had been pretty shocked at first by the wheelchair. C.J. had almost immediately given them the whole story—how he'd gotten into a bad auto accident while in high school, hit his head on the windshield and severed his vertebrae, leaving him a quadriplegic. Even so, he seemed to cope extremely well; he could speak, move his head a bit, and use his hands. As a software designer, he'd explained, he didn't need much else. Scott was extremely impressed by his resilience, and especially his optimism. He was all smiles behind his constant cigarettes, and with his mop of jet-black hair, his animated features, and his wide smile, you almost forgot about the chair.

His brother was much quieter—lean, tall, with narrow features and a sharp, almost beak-like nose. Not particularly friendly;

the only conversation Christian had initiated since they'd arrived at the offices was about the next fifty-thousand-dollar payment—and what it was going to take to get it sent as soon as possible. Scott had assured him that as soon as they put together a good beta version of the poker software, they'd get that check.

"And especially," C.J. continued as he used the keyboard to pull up what they'd been working on since getting the first payment, "since you're so set on poker. A sports book—now, that's something you could get in no time. That's where the money is."

There it was again, the same damn refrain. Scott vigorously shook his head.

"Poker."

"Poker," Garin added, almost simultaneously, as he leaned over the other side of the redwood table, looking toward the screen. "And like we said on the phone, real sophisticated-looking, James Bond kind of shit. Like, you order a martini, you smoke a cigar, and you play a little poker."

C.J. did his best to nod. He flicked a finger toward the screen, which had now lit up with a primitive-looking website. "This is just a mock-up. We still have many questions, and I do apologize, we're not yet as familiar with the game as we need to be."

Although the site was truly basic, Scott immediately recognized some of the design cues that he and his team had suggested to C.J. over the phone. The oval poker table in the center of the screen was a deep royal blue. The chairs looked like red velvet, and there was a little cityscape in the background—very cosmopolitan. It had a long way to go, but it was a thrill to see even the most basic elements up there on the screen.

As soon as the digital cards began being dealt, however, things went rapidly downhill. Everything seemed sluggish, as if the cards were floating through a thick soup. And the game play was just flat-out wrong. All the cards were being dealt faceup—so that they could see what was going on—but even from the start, Scott could tell that the Koreans had no knowledge of the game itself.

"I think we're going to have to go over the rudiments of Hold'em again—" Scott started, but Garin was a lot less subtle.

"Why are there five aces?"

"That too," Scott said. "Look, I'm not telling you anything that you don't already know, but poker is even more complicated than I think you realize. You need more than just software that can make transactions. With poker, it's all about game play, and there are a lot of important timing issues: when you put your bet in, how long it takes for the chips to get there, how long it takes to deal the cards, everyone getting info at the same time. And everyone's using a different type of connection—dial-up, DSL. And what do you do when someone loses their connection? What do you do when someone goes offline?"

Scott was rolling now, and the room had gone real quiet, everyone just watching him and smoking. He knew he wasn't telling them anything magical, but if he was going to pay these guys $150,000, they were going to have to indulge him.

"You play poker with a bunch of your friends, you expect shit to happen. Someone misdeals, someone spills a drink, someone gets upset and turns over the goddamn table. Over the Internet, everything has to work a certain way. There has to be a flow. And most important of all, nothing can interrupt the play. The game has to be available to anyone who wants it, whenever they want it. If this seems too complicated for you, if these problems seem too difficult—well, we can take our business elsewhere."

Scott hadn't meant to end his diatribe on such a hard note, but, well, there it was. This wasn't a game to him—this was his life.

If C.J. was put off by Scott's tone, he certainly didn't show it. In fact, his face seemed to light up behind his ever-present cigarette.

"Complicated doesn't have to mean impossible," he said. "In software, music, in life, it's the complications that make a thing worth doing."

That, and one hundred fifty grand, Scott thought. Even so, looking at that computer screen, seeing that blue table and those pixelated velvet chairs, it was hard to stay cynical.

After all, given enough time and guidance, a guy who could run a software company out of a wheelchair could certainly learn to play Texas Hold'em.

The girl was only five foot two, but at least half her height seemed to be legs. They were bare all the way to the thigh, her skin so smooth and tan and toned that it almost seemed shiny, and the rest of her draped in a white-on-white silk gown that shimmered around her long, lithe lines, revealing way more than it obscured—everything about her was damn near spectacular. And she was right up next to Scott on the leather banquette, close enough that he could smell her floral perfume, could see his own reflection in the glassy black strands of her shoulder-length hair. It took every ounce of his willpower not to reach out and touch her, but C.J., his wheelchair rolled up next to the low glass table in front of the banquette, working his way through his second pack of cigarettes, had been extremely clear. Unlike the Del Rey, this was a look, don't touch, kind of establishment.

Officially, C.J. had explained when the black Mercedes had deposited them at the front entrance, the place was a karaoke bar. A rectangular, warehouse-style building, it was outfitted almost entirely in leather and glass, with multiple levels connected by an open, ascending spiral staircase, attached to the underside of which was the most obtrusive speaker system Scott had ever seen—giant, conical woofers and subwoofers dangling like futuristic barnacles. Still, Scott didn't have to be a genius to realize that karaoke wasn't the place's main draw.

To be sure, there was a stage near the back of the giant hall, bordered on two sides by enormous projector screens. There were microphones set up, and a constant stream of inebriated Korean men stumbling up the four steps to that stage—taking turns at the mike, warbling incomprehensible words as images flashed across the screens—but the karaoke was really just a background screech.

Once they were seated at the banquette, first came the bottles of whiskey—brands Scott had never heard of, with price tags he couldn't believe. And then came the girls—brought to the table in groups of three and four by a hostess in a clingy blue dress. C.J. had explained, almost apologetically, that the girls were there to be looked at, to giggle at your jokes, and to pour the whiskey. If Scott and Garin were looking for more, C.J. assured them that he could take them to places with less restrictive menus.

Scott was happy right where they were. They weren't in Seoul to get laid; they were there to build a partnership. He was content to let the beautiful Korean girl sitting next to him pour his drinks. Besides, it was obvious that she didn't speak a word of English. Still, the way she was smiling at him, intermittently letting her hand brush against his thigh . . .

Scott turned away from the girl, forcing himself to concentrate on the two Korean software programmers on the other side of the banquette. C.J., in his wheelchair, up against the glass table, knocking back cigarettes and scotch. And Christian, next to Garin on the two-seater directly opposite Scott. Christian was in midsentence, again going on about the second fifty-thousand-dollar payment. Garin was assuring him that the money would be there when the site was ready. If it had been Christian alone, and a team of Koreans who couldn't speak English and didn't play poker, Scott wouldn't be going back to Seattle with any sense of confidence.

But looking at C.J., who seemed to be perfectly at home in the karaoke club, with the screeching singing in the background and the goddesses strolling past the banquettes, Scott felt much more at ease. His gaze drifted to C.J.'s hands, the way his fingers almost imperceptibly bounced up and down against the wheelchair's armrests in rhythm with the music.

Then he felt another brush of motion against his leg and turned to see the beautiful Korean girl holding yet another bottle of scotch. Her eyes were low, not meeting his, but again there was a smile pulling at the corners of her lips.

No common language, a culture so different she may as well have been an alien—Scott smiled right back at her.

Even if the software still had a long way to go, it was clear that he and C.J. at least agreed on one thing.

He let his hand glide along the banquette, let his fingers rest against the girl's bare thigh.

Complicated doesn't have to mean impossible . . .

CHAPTER 11

"Well, this is encouraging," Scott said as he climbed down from the rickety bus, staggered through a thick cloud of exhaust mixed with yellowish dust from the poorly paved road, and stepped up onto the sweltering curb. "Nothing says international banking mecca like a guy pissing in the street."

"At least he's aiming for the grate," Hilt responded, exiting the bus behind him. Hilt had his suit jacket off, his tie flung back over his shoulder, but still his white oxford shirt was nearly soaked through with sweat. Being from Florida, he hadn't uttered so much as a single complaint during the two-hour, un-air-conditioned bus ride from the tiny island airport, but Scott had to believe his stoic friend had suffered just as much as he had. "I think that kid sitting next to you on the bus was pissing right onto the floor."

"And could you blame him?"

The bus ride had been an ordeal, and not simply because of

the heat. Once they'd left the grounds of the small island airport, the road had been almost entirely unpaved, winding in and out of what appeared to be undeveloped jungle. Even as they'd entered Roseau, the island of Dominica's capital city, the ride hadn't gotten much smoother; the road remained unpaved, even as untouched jungle gave way to urban poverty, punctuated every now and then with a glimpse of the Caribbean. They knew from brochures that there were a handful of resorts on the other side of the city, situated on white-sand beaches, but the bus had given that part of the island a wide berth.

Scott and Hilt were the only two Americans taking public transportation, but everyone had been polite and friendly. Even the man now standing not ten feet away, urinating toward the sewage grate in the center of an intersection just beyond the bus stop, was smiling. In fact, the guy caught Scott's stare and paused long enough to offer him free advice.

"You can do it too!" he shouted in a heavy island accent.

Scott laughed as Hilt joined him on the curb, straightening his tie while peering past the pissing man to the small row of buildings behind him. By the time the city bus had pulled away, its balding tires kicking up a new cloud of dust and dirt, Hilt had found what he was looking for—pointing his free hand at what appeared to be a tiny, single-story building set behind a pair of palm trees. The place couldn't have been more than one or two rooms, with brick and cinder-block walls and only a tiny, barred window. There was no sign out front, no parking lot, not even really a driveway—just another strip of dirt leading up from what was supposed to pass as a road.

"That's it?" Scott said.

"You were expecting the Taj Mahal?"

Hilt pulled his suit jacket over his shoulders. Scott shook his head. A guy pissing in the street out front, a building smaller than your average Burger King—he'd set his expectations low, but this was kind of ridiculous. He nervously patted at his lapel, feeling the thick checkbook that was secure in the inside jacket pocket. To him, that checkbook may as well have been made out of solid gold. He could feel Hilt looking at him, and he knew that his friend understood exactly what he was thinking.

It wasn't just a checkbook; it represented six months of their lives, and more hard work than either of them had done in four years of college. They had sweated and bled—sometimes quite literally—for every penny in the temporary account associated with that book. Phone calls, letter-writing campaigns, flights to and from Florida, New York, Washington, D.C.—hours and hours chasing down possible leads, potential investors. Slowly building toward their goal. Once they'd hired the Korean software designers, it had been easier to start securing real commitments. But not a single dollar of the $750,000 had come easy.

Just a week earlier, minutes before a presentation to a group of wealthy real estate developers in northern Florida, Scott and Garin had actually come to physical blows, after Garin had inadvertently left a piece of their presentation in the hotel room. Hilt had been forced to jump in and physically restrain Scott before they completely trashed the rented conference room. They had still managed to finish the presentation, Garin's jacket ripped right down the middle, Scott's hair askew—but the flare-up of tempers had only proven to them how much they had all personally invested in this. It was a shared passion, and sometimes inspired people did stupid, stupid things.

Hilt finished with his jacket, then led Scott toward the small building. As Scott followed, he only hoped that this wasn't one of those stupid, stupid things. The island of Dominica was a speck in the Caribbean that none of them could have found without Google, and yet here they were, in the island's only real city, checkbook in pocket, potentially ready to place all the money they had raised into the hands of complete strangers.

When the Costa Rican lawyers who were handling their incorporation had suggested that they use the Caribbean bank, Scott had understood the logic behind the idea. Absolute Poker was going to be an international company, with worldwide clientele. And despite what every lawyer had told them, the fact was, American banking laws and practices seemed to shift week by week. But now that they were actually there, strolling up a dirt path toward an unassuming wooden door, it was a lot harder to stay logical and relaxed.

The lawyers had given them three island banks to choose from: the Loyal Bank of St. Vincent, Bank Crozier of St. Lucia, and Bank Caribe of the Commonwealth of Dominica. They'd immediately crossed off the Loyal, believing that if the company had to put *loyal* in its name, it probably wasn't. Which left Dominica and St. Lucia. They'd decided to check out Dominica first, for the simple reason that the flight from Seattle had been cheaper. They were going to head to St. Lucia the next day. In between, they'd be staying in a hotel just around the corner, a disgusting little place Hilt had found on the Internet that offered tiny un-air-conditioned rooms teeming with cockroaches—because every penny they had would be going into that check.

Hilt paused at the door, giving Scott a chance to fix his tie one last time. Then he led the way inside.

"I have to admit, this all looks pretty good."

Hilt was leaning forward in the seat next to Scott, at the edge of the mahogany desk, poking through the huge stack of papers in front of them. Balance sheets, financial statements, asset allocations—all of it printed out at their request by the bank manager, who had now stepped outside to give them time to look through things in private. Even more important than the papers, to Scott, was the manager himself; Scott hadn't been able to stifle his surprise when he first stepped foot into the island bank and caught sight of the well-dressed, midfifties American, with his neatly combed silver hair, traditional-looking wire-rimmed glasses, and impeccable gray suit. The man had been accompanied by his son, a younger version of himself, with similar glasses and a similar suit. They both seemed extremely sharp and knowledgeable about the banking structures for the gaming industry, and once they had all situated themselves in the manager's office, the two men had been very open about all aspects of their work. Bank Caribe handled many gaming sites, from sports books to online casinos, and they had even worked with Paradise Poker. They also handled money from many Wall Street firms, including dozens of American-run hedge funds. Everything the men told them seemed proper, professional, and satisfying.

Now, forty minutes later, it was time to make a decision.

"What do you think?" Hilt said, leaning back from the papers and looking at Scott. "Half here? And half in St. Lucia, if it seems equally professional?"

As usual, there was almost no inflection in Hilt's voice. To him, this was business, and now that he'd seen the papers, he was able to put his emotions aside. Scott wondered if he himself would ever be able to be that levelheaded. For the moment, though, as he reached for the checkbook, he could feel his fingers trembling in time with his rapidly increasing heartbeat.

"Call the manager back in. It's time we open our first real bank account."

Finally, Hilt cracked a smile.

"You think he'll offer us champagne, to take back to our lavish hotel suite?"

"For three hundred and fifty thousand dollars," Scott said, breathing hard, "I'm going to hold out for a goddamn toaster."

CHAPTER 12

"Oh, man, this is never going to work."

"Shut up. Act casual."

"Dude, abort. I'm telling you, this is gonna end bad."

Garin gave Shane a shove and a glare, then quickly pushed his oversize luggage cart a few feet ahead, following the long line of tourists toward the double doors at the edge of the baggage and customs area. Shane hissed more terrified warnings after him but slowed his pace, letting Garin drift a few places ahead in the moving line.

Scott wanted to laugh at his two friends' antics, but he was a bit nervous himself. Looking at Garin's luggage cart—weighed down by so many cardboard boxes that the little steel wheels of the cart were twisting and turning against the tiled floor, emitting odd squeals that seemed to echo off the walls—he found it hard to believe that he was going to make it through. Scott himself had three cardboard boxes on the cart he was sharing with

Hilt and Shane. But his boxes contained only assorted clothes, shoes, and a handful of bathroom products that he thought he might have trouble finding in Costa Rica. The cardboard boxes were a hell of a lot cheaper than a suitcase, and he'd be able to throw them out after he unpacked at the house—in case, like in his dad's basement, they eventually decided to turn the closets into bedrooms. Though if the house was anything like Shane had described it after he'd gone down, a few weeks earlier, with an IT guy to set up the phone lines and computer wires, well, Scott doubted they were going to have any issues with space.

But five yards ahead and closing fast on the double doors that led out into the airport proper, Garin's cardboard boxes weren't just full of clothes, even though the scrawl of Magic Marker across the cardboard seemed to indicate just that. In reality, the goofball had packed away his entire desktop computer—monitor, hard drive, modem, keyboard, even a pair of speakers—and then piled clothes on top. Nobody had asked him any questions when he'd checked the heavy boxes in at the airport in Seattle, but now that they had landed at Santamaría, well, God only knew what would happen if the customs agents—who were milling about behind a long metal table just ahead of the twin doors in a group of about nine, in full uniform, with sidearms strapped to their hips—decided to pull Garin out of the line. Was a desktop computer something you were supposed to declare?

Looking at the other people in line—almost all of them tourists, a mix of families, young couples on honeymoons, groups of guys on their way to party, golf, or fish—it was obvious that almost every American heading to Costa Rica was on vacation. Nobody else, as far as Scott could see, had packed up what constituted their entire lives—for what amounted to a one-way trip.

Scott had no idea how long they were going to stay, now that they were actually there. The house that Shane had found was rented for a year, at four thousand dollars a month. According to Shane, the living room was big enough for a half dozen cubicles, which he and the IT guy had set up after multiple trips to the Costa Rican equivalent of Home Depot. And supposedly, there was even a pool out back.

"Shit, I think the one with the mustache is going to nab him," Hilt hissed, interrupting Scott's thoughts, from a step behind.

Scott followed Hilt's gaze and saw the customs agent eyeing Garin as he approached the steel table. For a brief moment it looked like the agent was going to say something—but then he shifted his attention to a middle-aged man in a heavy down jacket, just ahead of Garin, signaling the man over to the table. Garin plodded nervously on. Even from that distance, Scott could see the sweat beading on the back of his friend's tautly muscled neck. Somehow he was still moving forward.

Just as he reached the double doors, he turned his head the slightest bit to give Shane a big smile—and a silent "Screw you." The screech of the cart's wheels still echoed off the walls.

Finding a taxi that could fit all four of them and Garin's computer was a true test of their patience; there had to be two dozen of those damn drivers grabbing at them as soon as they'd stepped out of the airport onto the sidewalk, some so aggressive and even threatening that Scott was worried one of his friends might start throwing punches. But eventually they were able to find one who spoke enough English to figure out that four athletically built guys and a half dozen cardboard boxes weren't going to fit in one of their compact little clown cars. For a handful of colóns—the Costa Rican currency, which was colorful, the wrong size, and looked like it had come out of a board game—he brought them to a friend with a van, who was nice enough to help them load the boxes into his trunk. Even so, by the time the boxes were all loaded and they'd jammed themselves into the backseat—literally on top of one another—they were all covered in sweat. The heat and humidity were intense, even though the driver had the van's air-conditioning turned all the way up and there was a pretty stiff breeze coming out of the hills beyond the city. But the heat was something Scott knew he would get used to; like the Spanish music that was now blaring out of the van's crappy speakers, so loud it made the windows shake, it was just another detail of their new environment.

Then they were off, the man driving like a maniac through the thick airport traffic on his way to the city proper.

"Give the man the address," Scott said to Shane, trying to find space in the back of the van between Garin's cartoonishly long arms and Hilt's spark-plug shoulders.

"There isn't an address," Shane responded.

"What the hell do you mean?"

"I mean in San José, nobody has an address."

Shane leaned forward so that the driver could hear.

"We're going to Escazú. Start with the Tony Roma's restaurant, go a hundred meters west, a hundred meters south. The white house on the corner."

Scott stared at him. "Really?"

"That's how it works here. You'll see."

The driver wasn't arguing, so obviously he didn't have a problem with the directions. Scott leaned back against the seat, trying to find a comfortable position. Outside, the city flashed by. It was midafternoon, just like the last time he'd arrived in San José, but the roads seemed even more congested, traffic going in every direction. As before, he was amazed at how alive the place felt, how it seemed to throb with energy. So many cars, people, noise—even above the insanely loud music pumping through the van's speakers, the sounds of the urban sprawl felt like a hand, reaching right through those trembling windows, grabbing at Scott's skin. This was it. He was really here; this was actually happening. It was goddamn exhilarating.

Nearly an hour later they were still engulfed by the thrill of it all when Shane shot a finger toward the window by his head.

"This is it. Casa Absolute Poker."

The house was situated on a low hill, in a pretty, leafy residential neighborhood in one of the more upscale suburbs of the city. From the outside it wasn't anywhere near as lavish as Scott's dad's house, but it was big, and there certainly was a pool. And the neighborhood was pretty nice—though there didn't seem to be any nearby transportation, supermarkets, or even a general store, and they didn't have a car. But as they exited the cab, the driver handed them a card, telling them they could call his company anytime. Scott tipped the man well—he had a feeling they were going to be taking a lot of cabs for the time being, whether they liked it or not.

Shane led them inside with a proud sweep of his hand. And once Scott was through the high Spanish door and into the living room, he had to admit that Shane had done pretty well for them.

The living room had high ceilings, a good number of windows, and plenty of natural light. The floors were carpeted, the walls bare, but the interior was exactly as Scott had hoped. Shane and the IT guy had put up a half dozen prefab cubicle walls, converting the room into a serviceable office space. There were wires running everywhere; bright orange and green rubber snakes, some as thin as Scott's fingers, others as thick as garden hoses, running between the cubicles, along the walls, even through the doorway into what looked to be a galley-style kitchen. It was an electrician's nightmare—but, as Shane had explained already, a truly necessary endeavor.

When Scott and his team had first been told about the house, they initially thought they could buy computer servers and build a small server room in the house's basement, or out back by the pool. But Shane and the IT guy had quickly discovered this wasn't an option. Within twenty-four hours of being in Costa Rica, Shane had found out that the Internet crashed at least once a day; likewise, power was something that could go off and on at random. So instead, Shane had located a professional third-party server hosting—a place with the fairly ridiculous name HostaRica. HostaRica was, it turned out, the country's largest hosting site, and Shane had been told it handled most of the gaming concerns in the area.

So instead of servers, Shane and the IT pro had spent their time setting up workstations for desktop computers, which would be Scott and his team's bread and butter. Each cubicle had been outfitted with enough wiring to handle a computer, a modem, and a separate phone line, and even though the Internet and the power would be going on and off at random, at least the servers running their software would presumably stay on.

Aside from the wiring, Shane had installed in each cubicle a basic desk and little shelving units. Five of the cubicles also had desk chairs, but the sixth contained just a desk and shelves. Scott had to smile as Garin headed straight for the chairless cubicle, lugging his boxed-up computer behind him. When Shane had been setting up the house, they demanded he call them in Seattle so they could approve any expenditures over a hundred dollars; the chairs, it turned out, were a hundred and fifty bucks a pop. The discussion had turned into a heated argument, with Garin finally exclaiming that they didn't need any goddamn chairs, he'd sit on a rock for all he cared.

So Shane had bought chairs for everyone—except Garin. But to his credit, Garin didn't voice a single complaint. He just went to work unboxing his computer and began the long process of wiring the damn thing back together.

Meanwhile, Scott continued his tour of the house.

The upstairs was as clean, sparse, and acceptable as the downstairs—more high ceilings, a pair of reasonably modern bathrooms, and five empty bedrooms. Shane had picked the biggest bedroom, which also turned out to be the only one with an air-conditioning unit built into the window. He'd also installed separate phone lines in all five bedrooms, which was a huge plus. In a pinch, they'd be able to use their bedrooms as offices as well, which meant they could double their staff without having to think about moving. Scott had no idea how fast the company could grow, but this was as good a first headquarters as he could have hoped for.

When he was finished with the tour, Scott gathered his team back in the living room. Hilt passed out beers from a cooler Shane had left in the kitchen, next to the refrigerator. Scott guessed the cooler had something to do with the shaky power grid, but he tried to put thoughts like that out of his head. There would be plenty of time to worry about such things in the weeks, months, and maybe years ahead.

Right now, it was time to get things started.

"Gents," he said, raising his beer, a long-necked bottle covered in Spanish writing. "This just might be the start of something beautiful. So let's set up some ground rules."

"Maybe no alcohol around the computers?" Garin said, eyeing Shane's beer bottle, which was resting on the edge of the cubicle wall closest to his newly unboxed monitor. "These wires look pretty shady to me. I feel like my testicles are shrinking just from being this close to them."

"Your testicles are less important than the wires," Shane shot back. "And *you* try and set up a computer network in a third-world country. Just be glad we have running water and the toilets work."

Scott shut them both up with a wave of his beer.

"Testicles aside, Garin makes a good point. Once the computers are set up, booze stays upstairs, outside, or in the kitchen. The living room is for work. And that's what we're here for—to work. Everyone needs to be at his station at seven A.M. Eventually, we'll have to work in shifts. Six P.M. here is nine A.M. in Korea, so we'll have a team designated to deal with the software people once we launch the beta test."

The guys all nodded. Scott pointed to a small wooden side table Shane had set up outside the cubicles, near the door. There was a phone jack next to the table, and a place to plug in a modem.

"For now, that's going to be our Customer Service Department. We'll take turns manning it. From what we gather, there are going to be a lot of complaints coming when we launch the beta, with the power going in and out and the Internet going down. So there will be a lot of angry e-mails and phone calls. Always be polite. Always be professional."

"And on that note," Hilt said, standing at the entrance to the cubicle he had chosen, closest to the stairs leading up to the bedrooms, "no girls are allowed in the house."

Garin immediately groaned, and Scott raised his eyebrows. Then he thought about it and realized Hilt was probably thinking smart, as usual. If they were going to be running a business there, they had to keep their professional and personal worlds as separate as possible. This wasn't the SAE house, after all.

"Hilt's right. No girls in the house."

"What about the pool?" Garin asked. "Can we keep girls in the pool?"

"The pool is part of the house."

"Even if they can hold their breath for a really, really, really long time?" Shane asked.

"Those are usually the best girls," Garin added. "The ones who know how to hold their breath—"

"Shut up," Scott said. "No fucking girls in the house."

Garin shrugged. Shane sighed. Scott smiled and held up his beer.

"So that's settled. Now let's toast—because goddamn it, we made it this far. And I can't freaking believe that we're actually here."

They all shouted and drank. Then Garin placed his empty

bottle gingerly on the floor beneath his computer desk and clapped his hands together.

"If we can't have girls in the house, get that cabby back over here."

"Why?" Scott asked.

"Because someone needs to take us out on the town. It's time to celebrate."

Scott grinned. Garin was absolutely right—that was something they could all agree on.

It was time to fucking celebrate.

CHAPTER 13

Should we do a countdown or something?" Shane said from inside his cubicle. Like the rest of them, he was hunched over his computer, eyes glued to his screen.

"Don't be stupid," Hilt responded from the cubicle to Shane's left. "This is just a beta test. It's mostly friends, family, and SAE alumni."

"Still, it's like, momentous. We're going live."

Scott stretched his neck. He was in the cubicle next to Hilt's, looking at his own computer. His fingers were poised over his keyboard, and he noticed the slightest anxious tremor in his pinkies.

"I wouldn't call it live," he said quietly. "It's more like artificial life. We're going to stay in total control of the beta, make sure it goes as smoothly as possible. This is a test run, but we want people to come back when we're ready to launch the real thing."

Artificial life—that seemed like the right term for it. It would be AbsolutePoker.com, with real money from real accounts, maybe fifty or sixty of them spread out, mostly around the United States—well, mostly in and around Missoula, Montana, with a few in St. Petersburg, Florida—but they were going to monitor every second of play. The Koreans had created a handful of employee accounts, numbered accounts so they could keep track of the game play. They would also be able to observe player information, how the shuffle was working, how smoothly the tables were being filled. And after the fact, they'd be able to review the cards and the play in a timed log. For security, the Koreans had built a delay into the review process, so you weren't actually seeing any cards until after each round was over; but later on, Scott and his team would be able to monitor the game play to observe where it needed to be improved and how things could be made smoother.

The Koreans had done a pretty good job so far. It had been six months since Scott and Garin met with them in Seoul and four weeks since he and the crew had moved into the house in Costa Rica. In that time, together with the Koreans, they'd worked nearly round the clock getting ready for this day. And for all of them, it had been a truly brutal routine. Rolling out of bed, heading right to their workstations. Plugging away until nightfall, when one or two of them would begin the second shift, working with C.J. and Christian to hammer out the software details.

The Koreans had nailed the visuals almost from the beginning, but the game play had taken a lot longer. The main problem, as Scott had pointed out when he first met with C.J. and Christian, was that the game had to be smooth—the players must be able to sit down and play wherever they were, whenever they wanted to, even though they could be signing in from cities all over the world, with different modem speeds, hard drives, and Internet providers. When you got down to it, the software was endlessly complicated—and Scott was sure there would be plenty of issues, especially in the beta phase. The superuser accounts would be helpful in calling out some of those issues, but there was also going to be a lot of feedback coming in from the players.

Which was why at the moment, Garin was sitting cross-legged on the floor in the area they'd designated as Customer Service. In front of him were a computer and a telephone. The 1-800 number that led to that phone had been inserted into the beta introduction page, so for the moment Garin was their call center, ready for action.

"Okay, I guess it's time," Scott said. He signaled Hilt, who began to type into his computer, communicating with the server hosts at HostaRica, then with the Koreans, who would be monitoring the beta test as well. As Hilt had pointed out, it really was mostly friends and family—though they had also sent out notices through a few of the more public poker forums, advertising themselves in a decidedly grassroots manner. No hype, no giant promises—for now, just an offer of good, sophisticated tournament play in what they liked to think of as a uniquely cosmopolitan environment.

As Scott launched the software on the computer in his cubicle and the screen instantly shifted to the AbsolutePoker.com beta site, he couldn't help but smile, bathed in the deep blue of the table that dominated the center of the screen. In the middle of the table, the Koreans had added the Absolute Poker logo, which was a red diamond. Scott thought it looked great, set against that cosmopolitan blue. The chairs around the table were still velvety, with little tables between them—after all, you needed someplace to set down your martini glass. The room itself was carpeted in gray, but eventually there would be many room choices. And the cityscape that greeted players when they logged on would change too; the key, Scott felt, was to keep the game modern and interesting, and that meant there would probably be constant changes to the look and feel of the site.

But again, the visuals would be secondary to the game play. Paradise Poker, to him, had always felt clunky and primitive. Absolute Poker was going to be something different.

"And go!" Hilt said.

And then it happened. Slowly, one by one, the seats at the table filled. The players were represented by cartoonish avatars—but that too would eventually change. Scott envisioned that people would be able to pick and choose their own avatars, and maybe one day even upload their own photos. But for the moment Scott didn't care about the avatars. He was watching the cards, because the deal had just begun—and it was fucking beautiful.

"It's working!" he shouted. "They're playing. Look, that guy just made a bet. And that guy is gonna call—"

And then, suddenly, Scott's screen went blank. He heard shouts from the other cubicles—and then, a second later, the lights in the house went out.

"Christ no!" Garin shouted. "The power just went down."

"Is it just the power or the Internet?" Shane asked.

"I think it's the Internet too," Hilt said, his voice tight. "Crap."

"What do we do?" Garin asked. "Is it just us, or is the game gonna be screwed up?"

"The game too," Shane said. "Because the way we have the

beta set up, we need to be connected for it to work. Garin, get on the phone with the server—"

"Wait," Hilt said, as suddenly the lights flashed back on. "We're back online."

And Scott could see that they were. His screen was back up, the blue table in front of him again—but the game play had frozen middeal, a card floating halfway across the table, like part of a magic trick gone bad. Scott leaned back in his chair, his face reddening. Inside, he was furious. This was ridiculous. So fucking unprofessional. If anyone had lost money because they went down, they would have to reimburse it. He was sure any minute now Garin's phone would start ringing with complaints, and the e-mails would be coming in.

Scott leaned forward, resting his elbows on his computer table, his head in his hands. *Christ, what a way to start.*

And then, just as he'd expected, Garin's phone blared to life, a metallic, noxious sound that seemed to reverberate through the whole house. Scott watched as Garin grabbed the receiver, cupping it to his ear.

"AbsolutePoker.com," Garin started. And then he stopped. "Hey, Scott," he called. "It's for you. It's Glenn."

Scott looked at Garin, surprised. He had expected it to be a complaining player; Glenn Dwyer was one of the company's first new hires, an SAE alum who'd been a top student in the frat. Glenn was a CPA and MBA who had specialized in accounting, so they'd set him up as their nominal president, based out of Los Angeles. Really, his job was to handle their accounting and officially manage the $750,000 they'd gotten from their investors. He didn't usually call the house; Scott called him whenever they needed access to the money or had to make a particularly large purchase.

Scott felt Hilt's eyes on him as he crossed to the phone. He took the receiver from Garin and pressed it against his ear.

"Hey, Glenn, now isn't a great time, we've got some fucked-up issues going on with the beta—"

"Scott, we've got a major problem."

Scott could tell immediately from Glenn's voice that it wasn't going to be a little issue. The guy was one of the most mild-mannered people Scott knew—and at the moment, his voice sounded almost frantic, at least an octave too high.

"What is it?"

"Well, see, it started last Thursday. I was catching up on the banking receipts, and there were a few things I needed clarified, so I called over to Caribe to check with them. And nobody answered the phone. I tried again all day Friday—and again, no answer. I figured, hey, it's the Caribbean, people are pretty laid-back in the Caribbean, I can wait until Monday—"

"Glenn," Scott interrupted, "what the hell are you trying to tell me?"

"Well, I called back today. A guy from Pricewaterhouse-Coopers answered. He told me that Caribe Bank has gone insolvent."

Scott's throat constricted. "What?"

"The bank, man. It went under. Just folded. I mean, it's gone. That money's all gone."

Scott couldn't feel the phone in his hand. He looked up and saw that Shane, Garin, and Hilt were all staring at him. His face had gone white.

"What happened?" Hilt asked.

"Caribe Bank went under."

The air in the room seemed to freeze, like a leather belt snapping tight.

Then Scott lurched forward and vomited all over the floor.

CHAPTER 14

Brent Beckley tossed the last three spoons into the large plastic shopping bag, ignoring the cacophonic clash of metal against metal, then twisted the bag shut and slung it over his right shoulder. The thing was heavy, bulging at the bottom, and there was a very good chance it was going to rip right open, spilling three years' worth of collected silverware all over his bare kitchen floor. But he didn't have much choice. The guy who'd come for the potted plants had taken his last box, so the bag would have to do. Besides, the silverware was pretty damn shitty; most of the forks were bent beyond use, and the knives were so dull they might as well have been spoons. But for three bucks, it really was a case of buyer beware. Besides, Brent had been exceedingly honest in his Craigslist ad; he'd described about everything in his apartment as junk—and yet still, there he was, bagging up the very last of it. Every last thing had sold, from the tattered couches to the soap dispenser from the bathroom. Who the hell bought a used soap dispenser?

Brent shook his head, then headed to the front door. Outside, he carefully placed the heavy plastic bag of silverware on the floor of the hallway, where the welcome mat used to sit, before he'd sold that too. He thought about leaving a note taped to the bag—then decided it wasn't necessary. Nobody was going to steal a plastic bag of bent forks, and the guy who'd paid for it wouldn't have any trouble finding it; there was literally nothing else there.

Brent took his apartment key out of his pocket and stuck it in the lock. He said a mental good-bye, then turned and headed for the stairs.

Twenty minutes later he was sitting in the front seat of his rust red Buick Century, hands on the steering wheel, staring straight ahead through the windshield. He wasn't thinking, exactly, more like counting ahead. Seconds, minutes, hours—a sort of psychological exercise to calm his rapidly unraveling nerves. It was a terrifying thing, saying good-bye to everything you knew, starting fresh. But it was also exciting, the kind of thing you wanted to contemplate and remember.

Unbelievably, it was finally, actually happening. He couldn't begin to count how many times he'd called his big brother over the past six months, begging Scott to hire him, to let him drop out of school and come down to Costa Rica to join the crew. Again and again, Scott had responded the same way: *Absolutely not.* Brent shouldn't have been surprised. Scott had nearly forced him, kicking and screaming, to go to college in the first place. Without Scott, he would have probably ended up selling weed to rebelling Mormon high school kids in Salt Lake City for the rest of his life—well, until he ended up in jail, or in hell.

Instead, he had finally made it to graduation from the University of Montana. He'd cut his hair, traded his hemp shirts for oxfords with matching ties. He'd actually changed so much, cleaned himself up so thoroughly, that he'd been elected president of the frat house for his senior year, following in Pete Barovich's shadow.

And then, less than a week ago, the phone had rung. This time it was Scott calling him—with a job offer.

"Director of customer service," Scott had said. "We can only start you at two thousand dollars a month, but you'll be running your own department. If things go well, there will be a lot of opportunity for forward motion."

Brent had nearly dropped the phone. Director of customer service. That sounded like a pretty big title. Running his own department right out of college? It sounded like an incredible opportunity. Then Scott had dropped a bombshell.

"We need you here tomorrow."

Brent had laughed, thinking it was a joke. He'd just barely graduated; he had an apartment, things, a car. But Scott was dead serious. Brent knew the company had just gone through a huge trauma involving a bank failure, but obviously Scott and his team had decided to power through, and they weren't wasting any time. But packing up his life in twenty-four hours, moving to a foreign country to live, perhaps for a long, long time?

"Scott, I don't even have a passport."

Scott had paused for less than a second.

"Okay, by the end of the week. You get here by Friday, or you're fired."

And with that, he'd disconnected. Brent knew immediately that his brother hadn't been joking. Scott could be intense, to the point of manic, and though he liked to play hard, he did not fool around when it came to something important. He wouldn't think twice about firing his own brother if he screwed up. Especially with whatever had gone down with the Caribbean bank—and what Scott and his team were trying to do to survive after a blow like that—Scott obviously wasn't playing games.

So Brent went right to Craigslist. He'd sold everything in his apartment—down to the goddamn silverware—and then told his landlord that he was leaving. The guy had tried to argue about a lease, but Brent was paid up through the month, and there was simply nothing else he could do about it. Then he'd gone into downtown Missoula to get an expedited passport.

And now, three days later, and just a few hours before his flight—Missoula to Minneapolis to Houston to Costa Rica—he had just one more thing to take care of before he was in the air.

He tapped his fingers on the steering wheel for a few more seconds, contemplating what lay ahead, step by step, then yanked the keys out of the ignition and stepped out onto the curb.

The sun was high above the tree-lined street, barely any breeze pulling at his white shirt or his loosely tied paisley tie, the day beginning to bake, but Brent didn't feel warm at all. He felt ready.

He walked to the back of his car and got down on both knees. Then he pulled a screwdriver out of his back pocket. It took a few minutes to get the license plate unscrewed, and then he was back on his feet.

He crossed the street and headed toward a two-story building with glass front doors and smoky picture windows. There was a uniformed guard just inside the doors, but Brent walked right by him, heading straight through the marble-floored lobby. He'd been in the bank a dozen times before—usually to fill out forms, manage student loans, once in a while to deposit checks. It was also where he received wired money from Scott's dad to help him with tuition, insurance, and, of course, car payments.

Brent took a deep breath, tasting the air-conditioned air, then headed straight to the nearest teller; behind the window, the woman looked to be in her midthirties, with short auburn hair and too much lipstick, wearing a stiffly tailored pantsuit. It was lucky that there was no line, because today Brent wasn't sure he'd have the patience to wait.

He reached the window and, without pause, placed his car keys and the license plate on the counter in front of the woman.

"What's this?" the woman asked.

"Your car. It's parked out front."

The woman stared at him. Her red lips opened, then closed. "Sorry?"

"You own it," Brent said. "Twenty-eight hundred dollars to go, and I can't pay it—so it's yours. It's out front."

"Sir—"

"Call it a voluntary repossession. Be careful, the ignition sticks a little bit. I'd love to stay and chat, but I don't have the time. I'm moving to Costa Rica."

Before the woman could respond, he turned and headed back toward the glass exit, smiling as he went.

My God my God my God this is awesome, Brent thought as he spilled out of Santamaría International and into the overwhelming chaos, noise, and heat. He felt himself pulled in every direction at once, with people jostling him as they looked for their rides, taxi drivers yanking at his sleeves, kids trying to sell him handmade trinkets, people whispering offers to him for everything from booze to drugs to girls—and then he broke free from the crowd for a moment and saw Scott and Hilt over on the other side of the curb, leaning against a guardrail and just watching him. Both had big grins on their faces.

Brent composed himself as best he could and strolled toward them, flattening his tie against the buttons of his shirt. His backpack—all he had brought with him, filled with the few items of clothing that he hadn't sold—was sticking to his back because of the heat and humidity, but he didn't care. Seeing Scott there, waiting for him, took all the worry away. He'd have followed his older brother to the ends of the earth—which, looking around, seemed to be exactly where he was.

"Welcome to Costa Rica," Scott said, leaping off the guardrail to shake his hand. "And I see you came dressed to impress."

"I figured you'd want me to hit the ground running," Brent responded, greeting Hilt as well. Then they led him to a taxi that was waiting a few yards down the curb. It was obvious Scott knew the driver, who greeted him with a smile and a few words of Spanish. Brent couldn't tell whether Scott understood the guy, but either way, a moment later the three of them crowded into the back, and the cab took off like a rocket, zero to warp speed in less than three seconds.

"I mean, I've almost been fired once already, and I haven't even started working."

Hilt laughed, glancing at Scott. "You almost fired your own brother?"

"Hell," Scott said as the taxi narrowly avoided a truck carrying what looked to be bushels of bananas, then took a curve so fast it felt like they were up on two wheels, "we were all almost out of a job a week ago. Half our money, gone in the blink of an eye. Only managed to save the other half by pulling most of it out of the bank in St. Lucia, before it went bust like the one in Dominica."

Brent had heard about that from Phil. He'd told Brent how Scott had narrowly avoided full ruin by getting about three hundred thousand dollars out of the St. Lucia bank—only to hear that the St. Lucia bank then went under as well, just a couple of hours after he'd succeeded with the transfer.

"Even so," Scott continued, "that was nearly the end. We almost packed everything up and went home."

"But you didn't," Brent said, pressing his face against the window to get a better glimpse of the low buildings flashing by.

"Nope," his brother said, his eyes narrow. "We did the fucking opposite. We went all in. Decided to launch the site anyway, and then hit the investors up again, as hard as we could. Put a whole new valuation on the company, brought in as many new shareholders as we could find."

Brent was impressed, though he knew his brother well enough to understand: Scott didn't give up—no matter how bad things might be, well, he'd been through worse.

"We've revalued the company at four million," Hilt explained. "And the investors pitched in another seven hundred fifty thousand, so we've got over a million to spend on advertising and development. So we're kicking things up."

"And the site?" Brent asked. He'd logged on to it every night, waiting for Scott to finally hire him. He loved the way it looked, and although he wasn't really a poker player like his brother, he enjoyed the game play. He mostly lost, but never much, maybe thirty or forty bucks in total.

"Starting small, but still, the money has started to trickle in. We've been registering about fifty players a day. Most of those are playing for fun, but about one in ten puts down a credit card to play with real money. It's adding up to about a hundred bucks a day in our rake, but it's a start. As long as it keeps the board of directors happy, it's all good."

Brent looked at his brother. "We have a board of directors?"

Scott pointed to himself and Hilt. "And Garin, Shane, Phil, and a bunch of Hilt's people back in Florida. We're doing everything as by-the-book as we can, even though there isn't much regulation from anyone outside the company. Hell, we'd love some regulation, but the industry seems to work this way, so we need to do these things ourselves."

Brent had just graduated college—he didn't know much about business or how a company like this was supposed to be structured. But he could tell Hilt and his brother took these things very seriously. Then again, they weren't much older than he, and they were building their company on what felt like the edge of the world.

Outside the window, the scene was becoming increasingly urban, with storefronts and boxy apartment buildings, traffic jams and blinking streetlights. This didn't look like the upscale residential suburb his brother had described on the phone.

"Are we heading to the house?" he asked. He really was eager to get started working. As Absolute Poker's new director of customer service, he had a whole department to run.

Scott winked at him, then pointed out the front windshield of the cab. "Not exactly."

Brent squinted against the sun and saw a huge pink hotel rising behind a crowded sidewalk. His gaze slid up the seven stories to the marquee at the top of the building.

"The Hotel Del Rey," he read cautiously. "Are you checking me into a hotel?"

Again Scott winked.

"Not exactly," he repeated.

Six hours later Brent opened his eyes just in time to see the sidewalk hurtling toward his face. He managed to get one hand out, catching himself inches before he hit, and rolled onto his shoulder, then over onto his back, his feet straight up in the air. Above his leather loafers, he could see that the sky had somehow gone black. He had no idea how it had switched from day to night in what felt like a few seconds; the last thing he remembered clearly was stepping into that damn pink hotel, his brother and Hilt urging him on from behind. He'd lost sight of them when the first girl grabbed him by the hand, right inside the Del Rey's entrance, and he hadn't seen either of them since. Twelve hours ago he was handing his car keys over to a bank teller in Missoula, Montana. Now he was on a Costa Rican sidewalk, the sky between his feet. All in all, it had been a pretty good day.

Then he heard laughter, and female voices speaking Spanish. He followed the sound and realized why he was lying on the sidewalk. He'd obviously fallen out of the taxi when one of the girls had opened the door. There were two of them, the girls: one tall with dark skin and incredibly long legs, made even longer by her six-inch see-through heels; the other only about five feet tall, with short blond hair and enormous breasts. The girl in the heels was paying thc taxi driver—out of Brent's wallet, he realized—while the other one leaned over Brent, poking at his chest.

"You fall down," she said in heavily accented English.

"I fall down," Brent responded. He let her pull him to his feet. He was wobbly, but to his surprise, he found that he could, in fact, stand. As the taxi pulled away from the curb, he took the shorter girl in one arm and the tall, dark-skinned girl in the other. He had no recollection of where or how he had met them, but he could see that they were beautiful, and that they liked to share. Maybe this sort of thing had happened to guys like Scott in college, but Brent had never been in this situation before. And he liked it.

"Is your house?" the shorter girl asked.

With some difficulty, Brent peered out through the darkness, making out what appeared to be a two-story home directly in front of them. It looked just like the house Scott and Hilt had described when they'd given him the directions he was to use with the taxi drivers, but still he was amazed that he had gotten there on his own. Well, he'd had a little help. He had no idea how much he was going to have to pay the girls, but the fact that the taller girl was still holding his wallet was a pretty bad sign.

"I have no idea," Brent said honestly. "I hope so."

He pulled the two girls tight to his sides and started up the hill to the house's front door.

The porch light came on as he got close, and before he could even knock, or find a doorbell, or just kick the damn thing in, the door swung inward. Garin was standing in the doorway, all ten feet of him, with a goofy smile on his face.

"Brent, man. You made it. Welcome to the house. Scott was about to send out the Costa Rican air force to find you."

Brent smiled back at him and started forward, but Garin held out a hand, blocking his way. "Sorry, man. You can come in, but the girls can't."

Brent stared at the hand. It looked ten times as large as it was supposed to. "What do you mean?"

"No girls allowed in the house. It's an official rule."

"But . . ." Brent started. He gestured at the tall girl, then the other. "Two of them, Garin. There are two of them."

"Yeah, I can see that. You'll have to call them a cab."

Brent couldn't believe what he was hearing, but Garin only shrugged. "Sorry, man. House rules."

It took another fifteen minutes for the taxi to return to collect the girls. By the time Brent finally made it inside the house, he was halfway to sober. Still a little angry, but clearheaded enough not to make a big deal out of it. If Scott could acquiesce to the no-girls rule, then he certainly could. He was Mormon, after all.

He took in the cubicles that sectioned up the living room, counting the wires and computers. It was an impressive sight. He spotted Shane at one of the cubicles, plugging away, his ear to a phone. Maybe he was on with Korea, or perhaps he was dealing with the server. A couple of the other computers were on and running, the screens filled with numbers. One of them, Brent knew, was keeping live track of people signing on to the site and especially new registrations. Every morning, Scott had told him, they looked at that number, and if it was high enough, they had a mini celebration. It was a sort of ritual—no matter how drunk they got the night before, the team checked the registration numbers together.

"This place looks great," Brent said. "So where's my department?"

Garin's smile got even wider. With a flourish, he pointed to a short side table right inside the door. There was a computer on it, and a phone right next to the monitor. Someone had placed a low stool in front of the keyboard, set almost all the way down to the floor.

Brent stared at the table, then back at Garin. "You've got to be kidding me."

"Welcome to the Customer Service Department."

Brent closed his eyes. The whole department was *him*. He didn't even have a desk. His chair was a stool.

"That computer has about two hundred e-mails on it, most of them complaints," Garin continued. "There will be a hundred more tomorrow, guaranteed."

Brent opened his eyes, sighing. Scott was probably getting a big laugh out of the situation, but Brent decided then and there he wasn't going to say a word; he was just going to take it in stride.

"So what do I do with those e-mails, exactly?"

"You answer them."

Garin patted him on the back, then headed for the kitchen. Drunk as he still was, Brent lowered himself onto the stool. He was determined to look on the bright side. He still was the director of customer service. Even if it was a department of one. He was the director of himself.

He couldn't help laughing as he powered up the computer and started answering those damn e-mails.

CHAPTER 15

Six months flashed by in a blur of eighteen-hour days: Brent chained to that coffee table; Scott, Garin, Shane, and Hilt camped out in their cubicles. Every morning they rolled out of bed, then gathered around Scott's screen to see how many new players had registered, how many new real-money accounts had been deposited. Then phone calls with potential investors, shareholders, board members. Reading through e-mails from the Koreans. Impromptu meetings about new promotional ideas—free roll tournaments, where players could join for no charge and play for cash prizes, advertising opportunities on various poker blogs, in online poker magazines, via e-mail blasts . . .

Every now and then, when someone realized the refrigerator was empty, straws were drawn to see who would make the five-mile walk to the nearest supermarket, lugging home energy drinks, rice, beans, pasta, and beer in oversize plastic bags.

And when there was something to celebrate—the thousandth account, the first day with three hundred dollars in rakes, a good review in a poker blog—late nights at the Del Rey, and then, when that started to get old, scouting trips to a half dozen other clubs, bars, and casinos that pockmarked downtown San José. Sometimes the party went all night long, ending in some hotel room with a half dozen naked *ticas*—Costa Rican girls, some of whom became girlfriends, others who were just there for the alcohol—and a buffet of other vices.

Even so, the focus was always on their work. No matter how late the party ran, there was an unwritten rule in the house—you were at your post the next morning, or you weren't getting paid. At first, the salaries were paltry—a couple thousand a month, barely enough to cover their lifestyle. As the site grew, and its valuation doubled again, those salaries began to change. Three thousand a month, then five thousand. Back in the States those numbers still wouldn't raise many eyebrows, but in Costa Rica the guys were fast on their way to becoming what all the *ticas* assumed they were—rich expat Americans, living the expat way.

Brent knew that something had changed, that the bootstrap feel of the company was being left behind, when Scott had a BMW 3 Series convertible imported—delivered directly to the house by a driver who actually asked if they were the legendary gringos who had founded Paradise Poker, telling him the same story Scott and the group had heard when they first toured the country about the shadowy foreigners who supposedly lived somewhere in the hills above San José and were shuttled around by bodyguards, though in this retelling it was Range Rovers with tinted windows instead of Escalades. When the BMW was followed by the purchase of two souped-up motorcycles, a Kawasaki Ninja for Shane and a bright red racing Ducati for Scott, it was just more evidence that things were building faster than any of them had expected.

Still, though the money was changing and the site was growing in popularity—getting more good reviews all over the Web for being clean, fast, and trustworthy, always paying its accounts quickly and honestly, even in an industry that was completely unregulated—Brent's life revolved around those e-mails. He got complaints about everything from Internet outages to slow deals, from nitpicky objections to the visual layout of a game table to reports of perceived abuse from other players. And, of course, numerous accusations of cheating by other players, by imagined robot players—artificially intelligent software that some magazine or another had suggested might be trolling the fledgling Internet sites—and sometimes even by the site itself. It was an early lesson for Brent, one that he assumed every casino operator in Vegas had learned the minute they offered their first game: when people lost money, their first instinct was always to suspect foul play. Game of chance, game of skill—it didn't matter; it was rare when players took losing gracefully. But Brent always did his best to address the complaints, no matter how arbitrary they seemed.

Even so, it was a mind-numbing, soul-crushing job. As more money came in, as Absolute Poker went from a fledgling little company with a handful of players to a viable business with enough investment to support a valuation nearing six million dollars, Brent decided it was time for a change.

Thankfully for him, he wasn't the only one to feel that way.

"And we are up and running!"

The applause was near deafening, accentuated by the low ceilings and the freshly painted, newly plastered walls. Someone popped a bottle of champagne, the cork flying in a low arc that narrowly sailed over the bank of cubicles and threaded between Scott and Hilt, who were standing at the head of the long, rectangular room. Behind them, a pair of windows looked out over a poorly lit parking lot. From where Brent was sitting, behind a computer screen at one of the cubicles, he could count at least twelve cars in the lot, including Scott's BMW, as well as two motorcycles—Shane's Ninja and a Honda owned by one of the directors of the sports book that took up the top two floors above them.

The building wasn't grand—it was really just a concrete, glass, and plaster box in a strip mall near downtown San José—but it was a hell of an improvement on the home office. Instead of seeming like a fraternity, the company now felt like a corporate business. Other than the dress code, that is; the Costa Rican employees, who made up two-thirds of the room, were mostly in short-sleeved shirts, shorts, and even flip-flops. Scott was dressed similarly; he'd taken to wearing flip-flops almost everywhere, even when they went out to fancy restaurants—which they did almost every Friday, now that things were going so well.

But even with the flip-flops, nobody would have mistaken the place for the SAE house anymore. There were at least thirty people moving about, engaged in some sort of job or another. Thirty people who were getting paychecks, having their health and dental paid for, setting up company trips to the beach, even taking part in the homeless charity—centered in a soup kitchen in one of the more depressing San José slums—that Brent had talked the other guys into helping him set up. They were a part of the community now, and Brent felt it was important to make sure their lives revolved around something other than printing money.

Which was exactly what it felt like Absolute Poker was doing, now that it was up and running. Every time they refreshed the screen that displayed new registrations, there were twenty or thirty more—and now a good portion of those included credit card payments, mostly through Neteller, one of the biggest online depositors around, and also PayPal, maybe the most respected online commerce facilitator on the Web. They were the real deal now, still lean compared with a handful of competitors that had moved into the market since they'd opened their doors—hell, it seemed like a new poker company was appearing out of thin air every second of the day—but making money hand over fist.

But despite their extreme success, their new digs were indicative of how professionally they were handling the cash flow. Other than the moderate salaries they were now paying themselves, almost all the money was flowing right back into the business, and nearly all of it was being spent on marketing. Because the most important thing they'd learned was that advertising and marketing were the lifeblood of their business. At the moment, Garin and Hilt were handling most of those concerns, but Brent knew that building a marketing department was just one of the main things on their wish list for the coming six months.

As for his own role in the company, things had finally shifted in a positive direction. He'd handed off the customer service job to a pair of Costa Rican employees with better-than-basic English skills and had moved into "fraud protection"—a kind of creative way of saying that his job was to monitor both the game play, looking for anomalies, and, more important, the financial flow coming in through the depositing agents. He was spending a lot of time on the phone with the online credit companies and the various big U.S. banks, making sure all their accounts were in order. Although there were a few banks that eventually decided not to accept transactions pegged toward gambling, most, including many of the biggest banks in the United States, were very happy for the business.

And why not? Business was good and getting better. Brent couldn't help but feel amused when he thought back to Pete Barovich's belief that nobody would be comfortable playing poker over the Internet for real money. The truth was, people were *begging* to play poker over the Internet for real money. And not just college kids, though they made up a huge part of their market. Adults were turning what started as a hobby into a profession; some of the bigger accounts they'd received were earning close to six figures through careful and skilled daily play. Some players were playing eight, ten hours a day—and earning hundreds of dollars a session. AbsolutePoker.com was making good money off that rake, but to Brent, it was also providing a market of sorts—really, directly akin to a stock market, or any other market that provided a place where someone with a lot of skill, and a little luck, could earn a good living.

Everything was coming together, better than Brent could

have imagined. As he knocked back his glass of champagne, letting the bubbles caterwaul down his throat, he locked eyes with his brother, still up at the front of the room next to Hilt. Scott looked tan and happy but certainly not content. But that had always been part of Scott's allure—he was never entirely content, because he was always driving forward. And always driving fast.

CHAPTER 16

The thick, humid night air whipped against the face shield of Scott's motorcycle helmet as he hunched forward over the slanted steering column of his bright red Ducati racing bike, trying desperately to keep the damn thing in the middle of the road. He didn't dare look at the speedometer; he could tell by the way the bike was trembling against his body that he'd passed seventy miles per hour when they'd hit the last straightaway, and there was a good chance they were way beyond that now. His headlight was little better than a flashlight at that speed, and against the inky black of the long, desolate stretch of blacktop that bisected what looked to be sugarcane fields on either side, it was almost as useless as shouting into the wind. Which he was doing anyway, though he knew there was no way in hell that Shane could hear him.

Shane was about a hundred yards ahead, streaking through the blackness on his Kawasaki Ninja, visible only by the tiny, jerking blur of his taillight. Scott was desperate to catch up to him, but even though the Ducati was a much more powerful bike, he didn't dare try to push it any harder. He was only thankful that they had chosen one of the few paved roads in the area; had they taken a left at the last turn instead of a right, they'd probably both be dead by now.

In retrospect, of course, neither one of them should ever have been in this situation. Going much too fast, driving recklessly—in Costa Rica for such a short time and already *living* recklessly. And with Shane, specifically, Scott and the rest should never have let it get to this point—and Scott definitely blamed himself. All the signs had been there, and anyone who wasn't blind or stupid should have been able to see them for what they were.

Sure, it had started simply enough, way back at the Del Rey that first time. Shane, drunk off his ass, pawing at the hooker strolling behind their blackjack table. The kid they'd all known as a straitlaced, under control—if a little obsessive—social star breaking character in the face of sudden, unregulated temptation. Looking back, of course that was the first sign, but hell, pretty soon they were all drinking just like he was; they were all grabbing cookies from the cookie jar until they couldn't eat any more.

When Shane had shifted from alcohol to weed, again nobody raised any eyebrows. They'd all smoked a bit in college, and now that it was basically a phone call away, there didn't seem to be anything wrong with a joint now and again. After all, they all worked so hard, and as long as everyone made it to their desks on time, who cared what they were smoking when they got off work?

But the cocaine—that was where it had all started to go wrong. It had begun real simple—in the back of a cab, asking the driver if he could get them some more weed. Well, what about coke? And the driver had simply smiled, pulling a plastic bag out of his glove compartment. Eventually, for Shane the taxis became a personal delivery service, bringing whatever he asked for, whenever he wanted.

In retrospect, all of them should have seen it happening. Shane's gradual deterioration, his leaving the bar a little earlier each night, having these strange side conversations with the taxi driver when he thought nobody was looking. Then, when the rest of them got back home at 3 A.M., Shane would still be wide awake, alone in his room with the air conditioner on full blast, talking a mile a minute to himself. Eventually, as he began showing up later and later at his computer station, he reacted angrily when someone pointed it out.

And after that—well, after that it just got weird. Scott would never forget the day they'd all come into the house to find a long blue cable stretching all the way from the power outlet in the basement, across the living room, up the stairs, then under Shane's door; he'd moved his computer station into his room so that he could stay there day and night. And that's exactly what he did. For weeks on end, nobody saw him. And when he did finally come downstairs, he looked worse than shit. Skinny, his hair falling out, his eyebrows completely gone. Had he plucked out the hairs in a neurotic coke haze, or had they fallen out naturally? Regardless, he was clearly out of control.

The final straw came about a week later, when Scott and the others officially moved out of the house—Scott, Brent, and Garin to the house Scott had rented in the hills, Hilt and his girlfriend into an upscale apartment in a gated complex near downtown. Everyone had just packed up and moved out. And then, a day later, Shane had finally wandered into the new office and asked if anyone had had a chance to pick up any groceries for the fridge. Everyone had just stared at him, shocked. They'd cleared the entire house out—even the furniture—and Shane hadn't even realized that they'd gone.

At that moment, every one of them knew that something had to be done. Scott knew he shouldn't have waited another minute—at the very least, he should have sat Shane down and said something. But instead he'd decided to give it another day to think the next step through.

And now here he was, screaming into his helmet as the three-hundred-pound aluminum-and-fiberglass beast between his knees fought to stay on the pavement, chasing a little red flash of taillight in what seemed to be infinite darkness. As far as he could tell, Shane hadn't been on drugs when he'd shown up at Scott's house, helmet under his arm, wearing nothing but shorts, a T-shirt, and a pair of flip-flops, revving the engine of his Ninja, asking if Scott wanted to join him for a ride. But even though he was sober at that moment, Shane was in the midst of a growing, dangerous addiction. Still, when Shane had donned his helmet—thank God—and taken off down the driveway, Scott had only grinned and torn right out after him.

And then, right there, as Scott leaned into a soft curve, it happened. One second he was looking at that taillight, red and wobbly, and then suddenly he was seeing Shane's headlight shoot straight up into the air. There was a terrifying squeal of metal against pavement, then a fountain of sparks sprayed out above where the headlight used to be.

Scott hunched farther forward and took the last fifty yards as fast as he dared. As he pumped the brakes, putting himself into a controlled skid, he caught a glimpse of the scene in front of him in the glow of his own light. The road had curved to the right—and Shane had been unable to take the turn or stop in time, hurtling directly into what appeared to be a highway construction site. A long braided wire hung at about waist level; Shane had managed to lay the bike down—and the Ninja had skidded out and gone under the wire. Shane had gone over.

It took Scott a full beat to see where Shane had landed—a good ten yards from where the bike had gone down. Shane was lying facedown on the dirt, his hands splayed out at his sides, not moving.

Scott cursed, tearing off his helmet as he leaped off the Ducati. He hurdled the braided wire and raced toward his friend. *Christ,* he thought, *he can't possibly have survived that.*

And just as he reached Shane's side, Shane rolled over onto his back and started trying to push himself up off the pavement. His T-shirt was shredded, and there was blood everywhere, spilling from gashes and cuts up and down his chest, arms, and bare legs.

"Dude," Scott gasped. He couldn't think of anything else to say. He looked at Shane's helmet and saw the huge crack across the front shield, running all the way to the back. But other than that, and the road rash up and down his body, Shane didn't look broken, at least not on the outside.

Before Scott could stop him, Shane yanked the helmet off. His eyes were glazed, but he was conscious. He tried to stand, but Scott stopped him with a hand.

"You're a fucking mess," Scott said. "You need to stay still."

Shane mumbled something—an apology, it sounded like, though he wasn't making much sense—and Scott told him to shut up and try to stop bleeding so much. At that, Shane cracked a little smile. Scott didn't know whether he wanted to punch his friend or grab him in a hug.

First he had to get him to an emergency room, to get X-rayed, scanned, and stitched up like a rag doll.

And then he had to get his friend the hell out of Costa Rica.

It was a hard phone call to make. Scott, Garin, Hilt, and Brent gathered around the receiver as one of them dialed, then they took turns with the phone, because at first Shane's mother refused to believe what they were telling her. They finally got her to understand how bad things had gotten. Since the motorcycle accident, Shane had grown even more isolated, the bandages that covered half his body a perfect excuse for him to lock himself in his room for days on end. His mother had immediately connected them to Shane's uncle, who was a rehab counselor and a former addict, clean and sober twenty years now.

Shane's uncle didn't ask any questions—he simply bought a ticket to Costa Rica and got on a plane the next morning. When Scott picked him up from the airport, Shane's uncle showed him a first-class ticket back to the United States with Shane's name on it; he wasn't going to leave without his nephew.

The intervention took place at the small apartment Shane had moved into after he'd finally realized that nobody else was living in the house anymore. They surprised him in the kitchen; when he eventually came out of his room, looking for something to eat, they were all gathered around his kitchen table. Although Shane's uncle led the conversation, by the end nearly all of them had gotten emotional; Shane had been there since the beginning, and it was crazy to think of the group without him. But he needed to get help, and he needed to get better.

By two in the morning, they were all helping him pack. When his uncle led him to a waiting cab, to take him to the airport for his flight back to the States, he had his head down, watching the ground beneath his feet. Hunched over, skinny, covered in bandages, eyebrows gone—he was a sobering sight for all of them.

But as the cab pulled away and they headed for their own cars, there was little talk of slowing down; if anything, Scott felt it was time to step things up. Shane had shown them—in the world they lived in, if you lost sight of where you were going, you ran the risk of ending up facedown in the road.

CHAPTER 17

Two months later, January 2005, and the party was just getting started.

It was the end of a regular workday. Brent followed the team back to Scott's newly rented house high in the hills above their former home office. Brent and Garin would be staying there as well, and although Brent had toured the place a few times since Scott first moved in, he was still awed by its scale. The building itself was massive, an old Spanish-style mansion with at least six bedrooms, a living room that could have doubled as a ballroom, and multiple decks overlooking downtown San José—a vista of sparkling lights, bolstered by the pulsing flare of the constant traffic that threaded between the buildings, like a radiated circulatory system feeding that ravenous urban sprawl.

The party was mostly confined to the house's massive pool deck. A DJ had set up shop on an elevated stage, the music from his spinning CDs blasting out of massive twin speakers that had been built right into one of the house's exterior brick walls. At five minutes to midnight, Scott strolled through the party, pulling everyone he could find out of whatever trouble they had gotten into to lead them to the farthest railing that looked out over the city. There was a Colombian girl on Scott's arm—Brent hadn't quite caught her name, but he thought it might be Clara. She was wearing a black bikini bottom and nothing else. There was a tattoo of a tiger on her lower back, and her dyed blond hair was tied back in an elaborate ponytail, held in place by a band that sparkled with what may very well have been diamonds.

By the time Scott made it to the railing, Brent, Garin, and Hilt were right behind him. Shane, just back from rehab, followed a few feet after, a bottle of water in his hand. Although Shane walked carefully, still on the road to a full physical recovery, the color was back in his cheeks; now that he had returned to the team, he was dividing his time between the office and NA and AA programs. A living, breathing, walking reminder of what life in a place like that could become if you let yourself lose control.

When the team had gathered around him, Scott leaned back against the railing, his arm around the topless girl's shoulders. Brent, Garin, Hilt, and Shane made a small semicircle, wondering why he'd brought them to the edge of the deck.

"You know all those stories we keep hearing about the crazy gringos who live up in the hills, living like rock stars, spending money like they're printing it themselves?"

He waited a beat, then grinned his trademark grin.

"From now on, those gringos are us."

And right on cue, the midnight sky above their heads exploded in a brilliant wash of fireworks, streaks of Technicolor sparks raining over the city below, so many explosions in such rapid succession, it seemed like the show could somehow go on forever.

CHAPTER 18

Ten hours in the air. Three more in a rented Mercedes convertible, going seventy miles per hour with the top down, buffeted by a fierce wind heavy with the scent of lush hills, ancient lakes, towering pines—the Pacific Northwest distilled and funneled through the senses.

By the time Scott found himself standing next to Hilt in the leather-paneled elevator, racing up the spine of a glass-and-steel office building in the heart of downtown Portland, he should have been exhausted. But in the months since Shane's rehab and return, Scott had been operating on less and less sleep. Every minute of the day had been dominated by the company; despite the initial skyrocket of growth that had put them on the map as one of the premier online poker sites, things had started to turn—and not for the better.

"We're simply not getting enough for our ad dollars," Hilt was saying, continuing the conversation that had carried them through much of the trip from San José. "I've done the calculations. With all the Web ads, the promotions through the poker blogs, the magazine pullouts—we're paying about three hundred fifty dollars per player. It's just not sustainable in the long run."

Scott grimaced, his palms feeling the cool leather of the elevator walls. He knew Hilt was right. Over the past few months, he'd grown much closer to his business-minded friend. Hilt had become more and more his consigliere—the guy he turned to first when he had questions about the business. Garin had been there from the beginning and was his oldest friend, but especially since their intervention with Shane, in Scott's opinion—though Garin probably felt quite differently about it—Garin seemed to have taken a step back. Maybe Scott was reading too much into things. After Shane's stint in rehab, there were plenty of discussions about being careful, maybe trying to attain a little more balance rather than continue with their full-throttle, business-first mentality. But Scott had only one setting, and balance had never been one of his strengths. And Hilt was fast becoming his day-to-day partner.

Another part of the shift had to do with girls. Hilt had recently coupled off with an old girlfriend—a spectacular five-foot-seven blonde who had graduated college in three years with honors, a scouted model who was now in school to be a teacher. She had come to Costa Rica so they could be together, and they were moving rapidly toward marriage. In contrast, Scott had fallen into a relationship of his own with a hot-blooded, exceedingly jealous Colombian he'd met at a club in San José. For whatever reason, Scott's girlfriend had taken an immediate dislike to the girl Garin was dating, a pretty Costa Rican, and instead had been pushing him to spend as much couple time as possible with Hilt and his likely future wife. Maybe Scott's girlfriend was hoping that Hilt's relationship stability would rub off on them; whatever the case, the fact that Scott's and Garin's women didn't get along meant less time spent together.

"And the landscape has changed," Hilt added. "So many more competitors now. Another poker site popping up every day. And they're passing us by."

That was an understatement; not only were there now hundreds, if not thousands, of online poker sites, but a handful of them were dwarfing AbsolutePoker.com. Specifically, PokerStars.com and PartyPoker.com were monsters in the industry. To be fair, they had both been well capitalized from the beginning; neither had started in some frat house in Montana. Party Poker's founders had their 1-900 sex line fortune and had rolled that cash into building what was now probably one of the biggest online companies, in terms of straight revenue, in existence. And PokerStars was following right behind.

Absolute Poker had certainly grown dramatically from its humble beginnings. Now valued at more than thirty million dollars, it was bringing in about eight thousand dollars a day in revenue. But with the competition, and the price they were paying to sign up new players, they needed to figure out how to leverage what they had, and expand exponentially.

Which was why they were in the elevator in Portland.

"Hopefully Greg can tell us what we're doing wrong," Scott responded as the elevator slowed at the fourteenth floor. "Because God knows, he's done everything right."

Scott had met Greg Pierson, CEO and founder of Ultimate-

Bet.com, one of their competitors—which was now valued in the hundreds of millions, one step away from the PokerStars and Party Pokers of the industry—when Greg had tried to sell him on a software partnership. Greg was a computer guy, a programming whiz who fell into online poker sideways; at the time Scott had first spoken to him, Greg's company, ieLogic, was creating skin sites for the gaming industry—basically, software that allowed instant, pop-up gaming on the Internet without downloads. Scott had remained in contact with the man and had always thought he was one of the smartest guys in the business. So when Greg had invited him to come by their offices in Portland, Scott figured it would be worthwhile to pick the man's brain. Ultimate Bet had streaked right past Absolute Poker, and Scott needed to know why.

The elevator doors opened, spilling them out into a brightly lit modern lobby, staffed by a woman in an impeccable suit seated behind a smoked-glass desk. After a brief wait they were led to a corner office with large picture windows overlooking downtown Portland. Pierson rose from his leather chair to greet them, and it was instantly obvious that things were going very well for the former software geek. He was a compact man in an off-brown suit, and his face was glowing beneath his crown of reddish hair, eyes beaming from behind his wire-rimmed glasses. He looked like he'd just won the lottery.

He ushered them to a pair of leather chairs behind a glass table. As Scott took his seat, he saw the Ultimate Bet logo embedded in the table, beneath a scrawl of carved writing: WORLD POKER TOUR CHARTER MEMBER.

Pierson noticed him reading the table and pointed with a thumb.

"Cost us a hundred thousand, and worth every penny."

"The table?" Hilt asked, incredulous. Pierson laughed.

"Fuck you. No, sponsoring the tour. Got us unlimited access to advertise on the TV show. Wednesday nights on Travel Channel."

Scott and Hilt nodded, but Pierson just looked at them, then shook his head.

"You guys don't get it, do you? *Wednesday nights on the Travel Channel.* That's national TV."

Scott smiled politely. He knew the show. It was one of a number of poker shows that had launched on television, including *World Series of Poker* on ESPN, which was probably the industry's best-known show. The idea of poker on TV had been a long time in coming, but it hadn't been until very recently that the game had finally made inroads into the medium, specifically because of a pretty interesting invention—the hole cam.

The problem with poker, as a visual sport, was that you couldn't normally see the players' cards—so it was hard to play along with them, because you were only seeing the results of the game after all the cards were laid down, rather than taking part in the action along the way. But the hole cam changed all that, by showing the players' cards as they were dealt, so that the viewer could witness the drama of the game as it unfolded. First instituted on a show in the United Kingdom, it had come to U.S. TV via both the ESPN *World Series of Poker* and Travel Channel's *World Poker Tour.* Scott hadn't realized that Pierson's Ultimate Bet had been a sponsor of the show—but at first, he didn't get the significance. Besides, they'd all sponsored all sorts of shit, from cruises to golf outings to charity tournaments.

"It's a fun show," Scott agreed. "We catch it now and then on the TVs in the sports book upstairs from our office."

Pierson pulled his desk chair around so he was sitting across from them at the table. He leaned toward them, looking like some sort of intense, bespectacled turtle.

"You're not getting what I'm saying. Scott, you know how much revenue we did last week? One hundred thousand a day. Every day."

Scott looked at Hilt. He could see the shock in his friend's eyes. Absolute Poker was doing eight thousand a day—and that was during a good week. A hundred thousand dollars a day? That seemed impossible.

"How?" was all Scott could manage.

Pierson grinned, obviously happy he had gotten through. "The game has changed, man."

"We know," Hilt started. "So many more online sites—"

"No. That's not what I mean. The *game* has changed. Poker. It's gone mainstream, in a major way. Because of shows like ours, and the World Series. It started with Moneymaker winning a bracelet, but it's gone so much farther now, it's straight-up mainstream. And that's what we've done. We've gone mainstream too."

Of course, Scott knew all about Chris Moneymaker, who'd won the *World Series of Poker*'s main event in 2003. Moneymaker had been a regular Joe, an accountant who won his spot in the event by playing a thirty-eight-dollar online tournament at PokerStars. When he won the World Series—going head to head against some of the greatest professional poker players of the era—his $2.5 million prize had inspired a whole new generation. Moneymaker hadn't trained in dark poker rooms or smoky casinos. He'd learned how to play at home, online. If a guy as average seeming as Chris Moneymaker could win the World Series of Poker, then anyone could.

"One in five Americans played poker last year," Pierson continued. "And those numbers are only growing. More people play poker than play baseball—it's our goddamn national sport. And now it's nationally televised. Travel Channel, ESPN, ABC, NBC."

Scott was starting to catch on. For the first time in history, the two-hundred-year-old game of poker was being beamed into living rooms across America. All over the world too.

Back in 1999, when Scott had discovered the game, poker had been relegated to the back rooms of bars like Stockman's and the frat houses and dorm rooms of college kids. Now it was on TV, right after the Yankees–Red Sox game or the Super Bowl.

"TV changed everything. TV is the key."

Pierson railed on, explaining that Ultimate Bet was advertising on as many shows as it could—and everyone else was doing the same thing. PokerStars had the *World Series of Poker* and the *North American Poker Tour* on ESPN, *National Heads-Up Poker Championship* and *Poker After Dark* on NBC. Party Poker was advertising everywhere, including sports events and shows that had nothing to do with cards.

And through television, these sites were working with the poker shows to publicize a whole roster of poker celebrities who were becoming as well known as professional athletes: Phil Hellmuth, Chris "Jesus" Ferguson, Doyle Brunson, Phil Ivey, Annie Duke. By sponsoring players, sites like Pierson's could further advertise themselves—through T-shirts, hats, whatever they could get the players to wear. They were after exposure, as much as they could buy—and from what Pierson was saying, because of TV, they could suddenly buy a whole hell of a lot more than Scott had ever thought possible.

As Pierson talked on, Scott was already trying to figure out a way to cut the meeting short and get the hell back on the road. He tried to signal Hilt with his eyes. He didn't want to waste another minute in that office.

Pierson was right. Television was a window into every American home. The faster Absolute Poker could get on TV, the faster it would end up in every American living room.

"Go, go, go!" Scott shouted, pushing Garin, Hilt, Shane, and Brent up the cement stairwell in front of him. "It's on in two minutes!"

Scott had just gotten the phone call from their Philadelphia-based advertising company a few minutes earlier: AbsolutePoker.com's first commercial was going to air during the last few seconds of *World Poker Tour*—in the same block as one of Pierson's own ads. Although Scott and his team had gone over the ad a dozen times since the Philly guys handed it in, there was something different about the idea of watching it live. They wanted to see it on the screen just like their customers would.

Scott and Hilt had come up with the concept of the ad during the drive back from Portland to Seattle. Just a thirty-second bump, a simple story that Scott felt captured the essence of what Absolute Poker was supposed to be. A guy at a computer in a well-appointed apartment, his friends in the next room getting ready to hit ladies' night at a local bar. The friends tell him it's time to go downtown—that the party is about to start. The guy declines; he says no, he's going to pass, he's playing online poker. Then the AbsolutePoker.com logo—and something like "Play it anytime, anywhere."

To Scott, it felt perfect. There was nothing seedy about it, nothing dark. Choosing to play online poker over going to a bar—it was just another form of entertainment, and hell, if you looked at the average credit card deposits of most players, it was an even cheaper, more responsible decision than heading off to ladies' night.

Scott made it to the top of the stairs and burst into the back room of the sports book office, right behind Garin, Shane, and Brent. The room was set up like a lounge, with couches, a refrigerator, and a handful of televisions hanging from three walls. The TVs were usually turned to sporting events, but Scott had called their upstairs neighbor right after he'd gotten off the phone with Philly; all the screens were now showing the *World Poker Tour,* which had just ended. The credits were rolling as Scott took his position a few feet from one of the televisions, his hand resting on Hilt's shoulder. Garin, Shane, and Brent were near a second TV, breathing hard.

And then, suddenly, there it was. The ad went by fast—thirty seconds felt like four heartbeats, maybe five—and then the show was over. Scott blinked, still seeing the Absolute Poker diamond emblazoned on his retinas. Then Hilt turned and raced back toward the stairwell.

"Back down," he said. "Let's see what that did."

The rest followed, nearly knocking him over as he pushed his way back onto the stairs.

A moment later they were in front of Scott's computer, looking at the back-end numbers that told them how many people were registering for the site in real time. Hilt was hitting the Refresh button with his index finger.

At first the numbers were the same as before—a few new registrations, followed by a few more. And then—*Christ.*

Twenty registrations to the website. All new players, all signing up to AbsolutePoker.com for the first time.

Then forty registrations.

Then eighty.

"How is this possible?" Garin whispered.

A hundred registrations. Then a hundred fifty. And it was still increasing, more and more registrations, every time Hilt hit Refresh.

"From one commercial?" Brent said. "Holy shit."

"If this holds up," Hilt commented quietly, "we're going to do twenty thousand in revenue in the next day."

Scott exhaled. From around eight thousand in revenue before the television ad to twenty thousand after. And what would happen when those television ads were everywhere? How many people would be coming to AbsolutePoker.com to play poker? How much money would they be able to generate?

"Is the software going to hold?" Garin asked. There was a tinge of panic in his voice.

Scott ignored him. He caught eyes with Hilt, and could see that Hilt felt it too. Something had changed, something huge.

Scott glanced behind them, and saw that most of their

other employees—maybe forty, fifty people—were now standing around watching the company officers. Scott reached out and pulled an office chair from in front of the nearest desk, then climbed up so that he was standing high above the room. He waited until the place had gone silent.

"Poker is everywhere," he said, hitting each word. "And from now on, we will be everywhere too. Television ads. And not just on poker shows. I'm talking sporting events, college basketball, college football. Wrestling—everywhere. We're not just an online poker company anymore—we're a marketing company. And the thing we're selling is the most popular game in the world."

The game of poker had changed, and Absolute Poker was going to change with it—hurtling into the mainstream.

Maybe Garin had his concerns, but Scott knew it wasn't the time to slow down.

It was their moment to step on the goddamn gas.

CHAPTER 19

The speedometer had to be wrong.

There was no way the BMW could be going that fast down a road this narrow, or one that was mostly dirt and gravel, with enough blind twists and jagged turns that it was way beyond serpentine. And yet there it was, those bright red digital numbers screaming out of the dashboard, telling Brent that he should be screaming as well. But instead he was laughing, so damn hard that the half bottle of tequila he'd drunk over the past three hours was threatening to tango back up his esophagus and drench the backseat of Scott's car.

That would serve his brother right. It was Scott who was behind the wheel, grinning back at Brent in the rearview mirror every time they took another hairpin turn and somehow stayed on the road, a wild, adrenaline-infused grin that was almost as terrifying as the numbers on the dashboard. When his brother got like this, anything could happen.

"Creo que está loco!" the girl seated on Brent's lap squealed—alerting Brent to the fact that there was a girl seated on his lap. In the same moment, he noticed that she wasn't wearing any pants, just bikini briefs and a T-shirt. Her ample breasts were fighting against the thin material of shirt, dime-size nipples jutting out as she took another swig from the bottle of tequila, which Brent now saw she'd brought with her from the bar. She was either terrified or excited, or both. Brent tried not to stare—because he could see now that his younger brother, Trent, jammed into the cramped backseat of the speeding BMW next to him and the girl, had gone bright red with the effort to do the same. Every now and then, however, the poor kid lost to the urge and snapped a quick glance at that heaving, perfect set.

Brent couldn't blame him. Because Trent really *was* still just a kid, even though he'd turned nineteen a few months earlier. A devout Mormon who'd just gotten back from missionary work in Chile, he wasn't used to Scott and his antics, and he'd certainly never been in a situation like this before.

Trent had arrived in Costa Rica only that afternoon, about six hours earlier. Just a three-day visit, on his way back to Salt Lake City. When he'd gotten to Scott's house, Brent had suggested they take him hiking in the jungle, or maybe to the beach. Scott had vetoed those ideas immediately: *He's just spent the past year preaching about Jesus. We need to take him downtown.*

And so they'd brought him to the worst place on earth, the same spot Scott and Hilt had taken Brent when he'd first arrived in San José. That pink hellhole full of girls.

To Trent's credit, he'd kept himself under control the whole time, refusing alcohol, turning away when the girls tried to throw themselves at him. At one point in the evening a Nicaraguan beauty had sat right on his lap in the bar, pulling up her tube top to reveal her naked brown chest. Trent had looked away until finally she shrugged and moved on to the next guy at the bar.

Brent wasn't sure, but it appeared the same Nicaraguan girl was actually now up in the front passenger seat, next to Scott, her long arms extended as she gripped the dashboard, brown fingers now turning white. At least it looked like her from the side, the same green tank. She was shouting something in Spanish too, probably about Scott's mental state, but Brent could hardly hear her over the sound of the gravel spitting up behind the tires. Girls like her, and the one on his lap—they'd been cycling through Scott's house at a pretty frantic pace, ever since the company had discovered TV and exploded. There was so much money coming in, it felt like there was a reason to celebrate every single night. Even though most of the twenty thousand dollars in revenue they were bringing in each day was going back into marketing, there was still enough left over to put them all in a different tax bracket—if Costa Rica had tax brackets. Brent wasn't sure what Scott, Hilt, Shane, or Garin were bringing in, but his salary had grown from his original two thousand dollars a month to closer to ten. It was more than enough to have imported his own BMW, which would have saved him from the terrifying roller coaster he now found himself on.

In the distance, out the windshield beyond the Nicaraguan girl's clawed grip on the dash, he saw another turn in the road, marked by a pair of palm trees. At first he felt a surge of relief, because he recognized the trees; they were only about a quarter mile from Scott's house. But then his eyes shifted down, to the road leading up to the trees—and his stomach went tight. There was something weird about the gravel there, the way it was glistening in the moonlight, and he realized that it wasn't just gravel, it was also sand—a lot of sand. Which, if the digital numbers on the dashboard really were real, and not just a figment of his imagination, would make it extremely hard, if not impossible, for Scott to make the turn without hitting—

Suddenly there was a scream and a horrible screech of metal and his whole world lurched up in the air, then flipped over, spinning and spinning, and then . . . nothing.

"You okay? You okay?"

Brent opened his eyes. He was looking at the sky. His body was pretty much horizontal, his legs were extended right out in front of him, and yet somehow he was moving, sliding along the gravel on his butt, and then he looked straight up and saw Scott staring down at him, eyes wild, blood caked on his lower lip, hair sticking straight up from his head like a demented halo.

"You okay? You okay?" Scott repeated.

Brent realized that Scott had him by his arms and was dragging him across the road. Brent looked back toward where he'd come from—and saw the BMW on its side, windows shattered, the whole damn front of the thing crushed in like an accordion. One of the palm trees was cracked in the middle, and the other was bent over, like it was bowing to the moon.

"Holy shit," Brent whispered. "Trent."

"He's good," Scott said, relief in his voice. "We're all good. Everybody's good."

They reached the side of the road, and Scott placed Brent

gingerly on the grass. Brent looked over and saw Trent, sitting there next to him. At first glance, Trent didn't look anything like *good*. He was holding a rag to his forehead, and there was blood everywhere, running down both of his cheeks, covering the front of his shirt, trickling all over the grass.

"Oh, man," Brent started, but Trent offered a weak smile.

"Just cut up a little," he managed. "The girls are okay too. But I think Scott's got a broken ankle."

Brent looked over to where Trent was pointing with his free hand. Scott was trying to calm the girls, who were about five yards away, also on the grass, jabbering in Spanish. Neither one of them looked hurt. When Scott left them to cross back to Brent and his brother, Brent saw that he was indeed limping pretty badly.

Scott sat on the grass next to Brent, reaching down to check out his ankle. It looked huge, swollen, and blue, and Scott grimaced as he gingerly removed his shoe.

There were suddenly sirens in the distance—still about ten minutes away, by Brent's guess. Scott heard them too, and seemed to forget about his ankle.

"I gotta get the fuck out of here," he said suddenly.

Brent stared at him. "What? No, man, you gotta get to a hospital. We all have to—"

"No, I gotta get out of here. I'm the CEO of a company. It's gonna look real bad if this gets in the press."

"You got in a car accident—"

"With a couple of hookers in the backseat. No, man, I gotta go."

Suddenly he was up on his good foot, and then he was hobbling away from the road, into the brush. One of the girls, the Nicaraguan, saw him going and for some reason decided to follow, jogging right past Brent and Trent. She put an arm around Scott and helped him forward, deeper into what appeared to be a field of waist-high bushes and weeds.

Brent watched them go. Then he turned back to Trent, who was also staring after Scott, awe in his eyes. The bleeding didn't look quite as bad as before, but he was definitely going to need stitches. The sirens were getting louder, and the other girl was still babbling in Spanish, but her voice was so high-pitched and she was speaking so fast, Brent had no idea what she was saying. He reached out and put a hand on his younger brother's leg.

"Trent, I'm so sorry, man." The kid had been in Costa Rica less than ten hours, and he'd nearly died.

Then Brent noticed that Trent was smiling.

"Sorry? This has been the best day of my life!"

Brent was on his way to the emergency room to get his brother stitched up when the police finally found Scott crouching in the field of weeds, the prostitute curled up next to him with the bottle of tequila. The officer had his handcuffs out and ready—until Scott opened up his wallet. Two thousand dollars American, and the officer quickly went back the way he had come, leaving Scott and the girl in the field, where Hilt eventually picked them up and drove them home.

CHAPTER 20

There was something uniquely pleasant about the feel of a good hammer against your palm. Maybe it was the weight of the thing, how it pulled at the muscles of your forearm with just the right amount of force. Or maybe it was how it looked, arced back above your shoulder, the steel head of the tool poised in the air with so much pregnant power, so much potential strength.

And then the hammer was flashing downward, tearing through the air in a perfect swing. It hit the cast dead center, spraying plaster all over the newspaper Scott had spread out across the floor of his home office to cover the Oriental rug he'd had imported the week before.

Without pause, Scott brought the hammer back up above his head—and then he heard a cough from the open doorway across the room.

"What the hell are you doing?"

He looked up and saw Garin standing there, mouth agape.

Garin had a laptop open in his right hand, the screen glowing with their revenue numbers for the previous week. Scott wasn't sure how long he'd been there, or even when he'd arrived at the house. Scott had been in the home office for most of the day, his cell phone off. It was a Friday afternoon, and he knew that a lot of their Costa Rican employees would be on their way to the beach, starting the weekend early. Scott had chosen to work from home, to keep from being distracted by the early exodus.

"This thing is driving me nuts," he responded. Then he brought the hammer down again, hitting the cast from the side. This time the cast cracked like the shell of a nut, and he laid the hammer on the newspaper and went to work with his fingers, prying the damn thing open.

"It's only been two weeks," Garin said, stepping into the room. "Your ankle is broken. Didn't the doctor say you're supposed to wear the cast for six?"

"Fuck him. He doesn't have to walk around in it. My ankle feels fine."

Scott got his foot free and tossed the remains of the cast into a trash can by his desk. Then he slowly rose to his feet, carefully putting weight on his injured leg. His ankle felt weak, but it didn't hurt.

"Is this *Apocalypse Now,* descent-into-madness kind of shit?"

At that, Scott had to smile. Before he could respond, Hilt pushed past Garin and into the room. He had an intense look on his face.

"Got another one. This time from an M&A firm in Canada."

Scott stretched his leg. Garin looked down at his laptop.

"How much?" Scott asked.

"Twenty-five million. I told them to go fuck themselves."

Garin looked up, his face blanching. Hilt laughed.

"I'm kidding. I told him we weren't looking for that sort of investment." Then to Scott. "We're not, are we?"

Scott shook his head. Still, it was pretty amazing to hear—yet another firm wanting to make an investment in their company. Even more amazing, he knew that the investment would just be a stepping-stone to an eventual IPO. Because these days that's all anyone was talking about in relation to Absolute Poker—the inevitable IPO.

They were now raking in seventy thousand dollars a day in revenue and signing up players at an amazing rate. And their timing couldn't have been better. The industry had completely exploded in the past month—beginning, of course, with Party Poker's IPO on the London Stock Exchange. That poker company, with the help of Morgan Stanley and DKW, had managed to pull off the largest IPO in the previous decade on the UK exchange. Its valuation had gone as high as thirteen billion dollars on the first day, ending at around nine billion. The founders of the company were now on the *Forbes* list.

Billionaires. Off of online poker.

And almost overnight, the industry had become real. The biggest banks in the world were calling everyone who had a piece of it to see how they could get involved. Major, respectable institutions, with teams of lawyers and auditors and CPAs.

Scott had taken many of the calls himself. He'd given the major firms access to their books, and everyone liked what they saw. One of the biggest banks, with headquarters in London, had gone into specifics; they wanted to take Absolute public—for no less than a billion dollars in an IPO. The only problem was, unlike Party Poker, which had started in such a large way that they'd been run with an IPO in mind from the very beginning, Absolute Poker had only two years of audited financials; to IPO, they'd need three at the minimum. They'd employed some of the most respectable big banks and auditing firms available, had engaged numerous high-priced legal teams, and had meticulously documented their financials. But they still needed one more year of records.

One more year, and they'd be worth a billion dollars. No problem, Scott had told the London bankers. We'll get one more year on the books. By the end of 2006, they'd be ready to go big—Party Poker big.

So this, now, was their trajectory. If all went well, in one more year they would all be on their way to becoming billionaires.

Between now and then they just had to keep their heads down, keep the revenues flowing, and turn down all the investors who were trying to buy their way into an industry on steroids. But Scott felt they also had to make a few changes—because now that the industry was legitimate and first class, they needed to get respectable—first world.

"It's not money we need," Scott responded. "We're in the due-diligence phase now. We need to look clean and pretty. Look at Ultimate Bet—those offices in Portland were top-notch. We need to look like that."

"You mean we need to get out of Costa Rica," Hilt said. "Open another headquarters somewhere respectable, where we can hire professional talent and make sure the big banks continue to take us seriously."

Scott nodded. He had been going through this with Hilt for a while, and they had already chosen a place they felt was right for them: Vancouver, Canada. First world, close to the United States, relatively cheap, and with highly educated talent that would look very good to the major European banks and exchanges. They'd even come up with who they thought they might bring in to help open the new office—someone who knew their business intimately and also had a strong background in marketing. The current conversation wasn't spontaneous—it was really for Garin's benefit.

Because Scott and Hilt had also recently come to another conclusion: they needed to make some other changes as well, for similar, somewhat cosmetic reasons.

"Along those lines," Scott said tentatively, "we need to make some slight changes involving the shareholder structures. Specifically, we've decided to tweak the board of directors."

At this, Garin paused, shutting his laptop. He looked from Scott to Hilt.

"What do you mean, 'tweak'?"

"Gray it up," Scott said. "Put a little age into it, because the bankers who look this shit over don't want to see a bunch of twenty-five-year-olds running a billion-dollar company. Some of us can stay on, but we need to switch some out to add some of our older investors."

Garin's face was reddening, and Scott could tell he was getting angry. Scott didn't want this to get personal; it was just business, and good business at that.

"We think that since you're more focused on the day-to-day—the TV spending and marketing, stuff like that—it makes more sense for you to step back."

There, he'd said it. Like pulling off a Band-Aid. Hilt had

remained silent throughout the conversation. Still, Garin kept glancing his way. For a brief moment, Scott thought that something was going to happen—something ugly. But then Garin seemed to slump his shoulders, just the littlest bit.

"I assume you two will stay on the board. And your dad?"

Scott nodded. His dad was good for the image of the company, as a high-powered financial player in Seattle. And he and Hilt—well, they really were running the show. And Shane, now that he was clean, would also remain on the board; his family had put up a significant portion of the money that had kept them afloat, making them and him major shareholders. But the rest of the board would be older, respectable shareholders.

Garin was obviously angered by the change, but in the end he'd have to accept things as they were. A battle right now would be foolish; this was a train that none of them wanted to get off.

Scott turned away, so that he didn't have to look at the cauliflowers of red still spreading across Garin's cheeks. He forced his mind to whirl forward, away from the tension in the room. He had other, more important things to focus on.

Billionaires. Was it really possible? Internet billionaires—the idea seemed hard to fathom. But if things continued the way they were going, it was inevitable—Scott would soon be nearly a billionaire. A nobody kid from a trailer park in Montana, who'd barely survived childhood—on his way to becoming a billionaire.

All they needed to do was get clean and pretty—and somehow stay that way until the banks were ready to take them public.

CHAPTER 21

It was definitely the most difficult pitch Pete Barovich had ever had to make.

He was sitting in the sunlit, tastefully modern kitchen of his brand-new home in Phoenix, Arizona. His hands were cupped around a mug of fresh coffee, and his two Labrador retrievers were curled up at his feet. The dogs were almost as new as the house, both jumbles of puppy energy; even though they'd spent the whole morning running around the construction site that was the backyard—new pool, new Jacuzzi, a deck for barbecue parties, even an outdoor wet bar—they were taking turns pawing at the laces of Pete's shoes. But Pete wasn't focused on the dogs or the coffee; across the table, his wife, Brandi, was sipping from her own mug, watching him intensely. She'd already guessed the theme of what he was going to say, but she was going to make him say it anyway.

"We have to do this," he finally opened.

It wasn't the strongest opening, but this was an unusual situation. Pete had been honing his marketing skills since his fraternity days; he'd become a true expert at selling, not just objects or ideas but the big picture. With Brandi it was different, because with Brandi he had only one option, and that was to be utterly truthful. It was like playing poker if your only move was to immediately lay down all of your cards.

"Fine," Brandi responded, brushing long strands of her brownish-blond hair out of her eyes. At twenty-six, she was a staggeringly pretty woman—a former pageant girl who'd made it all the way to Miss Montana, she should have been way out of Pete's league. More evidence that even back in college, where he'd first met her, he could market snow to Eskimos. "What is it, exactly, that we have to do?"

Yes, she was going to make him spell it out, every word. Because really, it was so damn insane. They'd literally just moved into their new house two days earlier. Adopted the two Labs. Started construction on the pool, as well as the refinishing of the three bathrooms in the gorgeous split-level ranch-style home.

"We have to sell everything. Take a massive pay cut and move out of the country. And we have to do it right away."

Pete laid it out for Brandi, step by step; the bottom line was, he had been dead wrong. He had underestimated Scott and his idea. He had thought poker couldn't work over the Internet—and he had been wrong.

Absolute Poker's biggest competitors were IPO'ing in the billions. Absolute Poker was now generating close to a hundred thousand dollars a day, and had raised more than fifteen million in working capital. The company was being groomed for a billion-dollar IPO by prestigious banks and hedge funds, who were all hoping to get in on the ground floor of an industry that had taken the financial world by storm.

When Scott had called Pete the night before, offering him a VP job—and explained that they were also in the process of hiring a Canadian CFO and setting up a respectable office in Vancouver—Pete had been forced to admit to himself, and to his friend, how wrong he'd been. To Scott's credit, he hadn't rubbed it in at all; to the contrary, he'd explained that Pete was an important piece in the puzzle, that they needed a guy like him to help get the company to the next step. Then again, Scott had always been very good at telling people what they wanted to hear—and at making people see his way. In Pete's mind, that's really why Scott had succeeded, why he'd been able to build his company from nothing and, in a way, help invent an entire industry out of something that Pete had thought had no potential. Maybe Pete could sell snow to Eskimos—but Scott could sell Scott to anyone.

However it had happened, now Absolute Poker was on an IPO path; all the banks needed were one more year of audited financials and a real, first-world headquarters—with an outstanding accounting department and a first-class CFO. The fact that Scott had turned to Pete to help build the Vancouver office, even after Pete had turned him down numerous times when the company was nothing but a pipe dream—it was humbling.

Pete would have accepted the offer right on the phone, if he hadn't had one last sale to make before he could pack up his suitcase and get on a plane.

"I'm not going to tell you that it's not crazy," he continued to

Brandi, while the dogs chewed at his laces. "Because it's completely insane. But it's also an opportunity we may never have again."

That afternoon Pete and his wife called in an ad to the local paper—moving sale, everything must go. At five thirty in the morning the day after the ad ran, there was a line of fifty people waiting by their garage door. By noon, people were buying the salsa out of the refrigerator, and Pete was trying to figure out how to ship two rambunctious Labradors and an angry wife to Vancouver, Canada.

CHAPTER 22

Three thousand miles, eight days. One hundred twenty cars racing from London to Vienna to Budapest to Belgrade, then everyone loaded onto airplanes and set back down in Phuket, Thailand. Racing again, Phuket to Bangkok, then back on the planes to fly halfway around the world to Salt Lake City. Back on the road from Salt Lake City to Vegas. A concert at the Hard Rock by the rapper Snoop Dogg, then on to Los Angeles, right down the middle of Rodeo Drive—and a wrap party at the Playboy Mansion.

It was the ultimate fantasy, yet it was entirely real; the Gumball 3000 wasn't a race so much as it was a party on wheels, attended by royalty, the ultra-wealthy, movie stars, rock stars—and one group of frat brothers from Montana in AbsolutePoker.com blue racing jumpsuits, flanked by a dozen girls in matching blue tube tops.

Six hours into the wrap party, Scott found himself standing

next to his father on the back lawn of the Playboy Mansion. He was wearing sunglasses, though it was well past midnight, and he was shivering, despite the temperature being in the midsixties. Then again, he was soaking wet; his rented tuxedo clung to his legs and torso, and his cummerbund and bow tie had gone missing. If he had to guess, he'd say the bow tie was floating somewhere in the infamous grotto. When he and Phil had stumbled into the rock-walled cove, he'd recognized the place immediately from the magazine. And just like in the magazine, the grotto had been filled with girls, many of them actual Playmates, all of them topless. Scott had done the only sensible thing—he'd jumped right in, tuxedo and all. Phil had followed. He'd only wished Hilt could have been there too, but Hilt was in the UK, meeting with investment bankers.

So now Scott was soaking wet, but nobody at the party was going to care; it was the Playboy Mansion. And at the moment, Scott had the run of the place. He knew that parked somewhere out front was an F430 Spider, gunmetal gray—the Absolute Poker logo painted across the hood, that AP diamond in the center clinging to the car's perfectly precise curves. He knew that the Ferrari, on its own, would be a spectacular sight; and it was next to a row of some of the most expensive, unusual cars in the world. A Lamborghini Gallardo, a Rolls-Royce Phantom, even a pink Range Rover.

Hef himself, and his bevy of girlfriends, had greeted Scott and his friends at the front door. As premier sponsors of the Gumball, they were at the top of the food chain. Hell, in Vegas, Scott had signed a dozen autographs for being CEO of what was now on its way to being one of the biggest companies on the Internet.

"Soak it in," he said, really to himself as they watched a group of Playmates wander toward them from the direction of the miniature zoo Hef kept on the premises. "A little while longer, and we'll all be living like this—"

He was interrupted by a commotion from behind, and he turned in time to see Garin being half led, half carried from the direction of the grotto. Garin had a foolish grin on his lips and a stream of blood coming from his scalp. One of the security guards who was helping him along was holding what looked to be a bikini top against the wound, trying to stifle the bleeding.

"What the hell?" Scott said as Garin was ushered past.

Garin looked back over his shoulder.

"I guess there's a no-diving policy. There should be a freaking sign."

And then he was gone, on his way to the front exit.

Phil was laughing hysterically, and Scott was just shaking his head. They'd collect Garin on the way out; he'd just have to do his best to keep from bleeding to death, because this party was going to go all night.

CHAPTER 23

Now, that's what I call service," Brent said as he stood next to Hilt in the entrance hall to their lavish two-room suite, staring at the man in the doorway. The man was wearing a gray-on-gray uniform with too many buttons to count and white gloves, and he was holding Brent's perfectly ironed tuxedo shirt out in front of him like it was the flowing cape of a matador. "I called down to ask for an iron, and they send me a guy instead who irons my shirt in five minutes flat."

The man bowed, handing over the shirt, and Brent fished in his pockets for a handful of euros. The man bowed again, taking the tip, and Brent shut the door after him.

"I mean really," Brent continued as he went back to the floor-length mirror on the wall by one of the room's many closets. "You don't get service like that in Costa Rica. Or Montana, for that matter."

"In Costa Rica—" Hilt started.

"The guy would have blown me for that many euros," Brent finished for him. "Fair enough. But he would have done a shit job ironing my shirt."

Hilt laughed, going to work on his bow tie. His tuxedo had come out of his suitcase with the creases perfect and the lapels straight. Brent had no idea how Hilt had managed the feat, especially considering the distance they'd just traveled. Costa Rica to London to Nice, and then, of course, the helicopter. Brent had never been on a helicopter before. The rotors were surprisingly loud, even through the cushioned headphones he'd been given by the pilot. Still, there was something so incredibly upscale about a helicopter trip; he couldn't help thinking the whole time, *This is how rich people live.* It didn't hurt that the view down through the bubble-glass windows was one of the most spectacular on earth: the pristine beaches, twisting roadways, anachronistic castles, and craggy hills of southern France flashing by at 150 miles per hour. Brent would have asked the pilot to slow down if they hadn't already been late for the party.

Nowadays, it seemed like they were always late for a party. Probably because there seemed to always *be* a party. The Gumball 3000 and the Playboy Mansion had been just the beginning; now that they were throwing themselves heavily into high-profile sponsorships and events, they were charting up the frequent-flier miles on a near-daily basis. Celebrity golf tournaments, major rock concerts, television specials, and, of course, huge poker events, all over the world. Brent had seen places he'd only read about before—Paris, London, Tokyo, Hong Kong—and now Monte Carlo, which from the air had looked like something out of a Disneyfied fairy tale, a gilded little enclave of castles and ornate mansions dug right into the top of a mountain. The cab ride from the helicopter landing pad up to their hotel only amplified the fantasy, the taxi twisting up that serpent's tail of a road. It was not lost on Brent that once a year these same roads hosted the premier Formula 1 race—the Grand Prix—which began and ended right in front of their destination, the famous Hôtel De Paris.

Lavish did not begin to express what the hotel was like. Situated across a manicured plaza from the famed Monte Carlo Casino, the 150-year-old building was a palace of jutting domes and spires, elegant marble columns, arched windows, and ornate stone carvings; a veritable army of bellmen met them at the front steps, ushering them past the statue of a mounted Louis XIV, through the cavernous lobby, and beneath the massive chandelier that hung from the domed skylight, crystal tentacles painting the very air in strokes of reflected light.

Their suite was only slightly less impressive than the lobby: antique furniture, canopied beds, a miniature version of the crystal chandelier hanging from the arched living room ceiling, a balcony with a view of the courtyard and the equally palatial casino. And the party they had come halfway around the world to attend was presumably already in full swing in one of the many private gambling parlors—swank caverns of velvet and marble.

Fitting, that the event was being held in the most famous casino in the world—since it was sponsored by Neteller, the now billion-dollar Internet payment processor that had become the de facto financial pillar of the online gambling community. Nearly three-quarters of AbsolutePoker.com's deposits went through Neteller, and Brent guessed the numbers were similar for all of its competitors—certainly Party Poker, PokerStars, Full Tilt, and Greg Pierson's Ultimate Bet, which had become the big four in terms of revenue, with AP following in fifth.

"If anybody should be blowing us," Brent continued, taking his shirt into the marble-lined bathroom just beyond the entrance to the suite—one of two similar bathrooms, each nearly as big as Brent's entire old apartment back in Montana, with a soaking tub the size of his first car—"it's the guys from Neteller. They did hundreds of millions of dollars this year, a lot of it facilitating online poker deposits. It's no wonder they sprang for a helicopter. If it wasn't for us, they'd be hustling porn like the rest of the Internet."

Brent exchanged the collared shirt he had worn on the trip from San José for the tux shirt, watching himself in a backlit mirror as he went to work on the buttons. He noticed that the floor was particularly warm beneath his bare feet. Heated marble tiles, he realized. They really hadn't spared any expense. He shouldn't have been surprised. Neteller was a class act—very different from most of the other payment processors Absolute Poker did business with when Neteller wasn't available to handle a percentage of their accounts.

It was funny—a year before, Brent had never heard of Neteller, or of any payment processors—hell, he hadn't even known that there *were* businesses that processed credit card payments over the Internet. But now he was fast becoming an expert in the financial gymnastics of the online world.

After he'd shifted from customer service to fraud and bank security, he'd assumed that his job title would remain stable for the foreseeable future; when one of the banking guys walked into his office not three months ago and placed a stack of papers on his desk, he had thought it was just more financials to look over for instances of potential credit card fraud. But then the guy had asked him if he'd be interested in yet another job change; he and Scott wanted to know if Brent would want to be responsible for payment processing—essentially, handling all aspects of the movement of money in and out of the company from players, via the middlemen processors such as Neteller and PayPal.

Brent had jumped at the opportunity. It seemed like a great area for growth, and a chance to learn about the inner workings of a business that was integral to how online commerce functioned. On his very first day, looking through those papers, he realized that Neteller had been overcharging them on transactions; they'd been taking 8 percent off all deposits to the company, when the number should have been closer to 4. One phone call, and Neteller had immediately handed AP a three-hundred-thousand-dollar credit. It had been just that easy—there was so much damn money coming through the business, nobody was going to haggle over a few hundred thousand here or there.

The deeper Brent dug into the processing world, the more he realized that the class acts such as Neteller and PayPal were an exception to the rule. Most of the other processing companies were extremely shady—appearing and disappearing overnight, often incorporated out of island territories that Brent hadn't even heard of and run through banks he wasn't even sure existed; these were companies you gave your credit card numbers to at your own risk. But sometimes they were also necessary; the flow of player money moving in and out of the poker site was constant and hungry, and the middlemen made that flow possible.

Of course, it was only Brent's opinion, but some of what these processors did struck him as taking place in a bit of a gray area—shady, if not outright illegal. As some bigger American banks enacted in-house rules governing the use of their credit cards for online gambling—even though there was no clear U.S. law concerning Internet gaming, other than sports betting—some of these middleman processors set up accounts that effectively hid where the money was actually going. Instead of online poker, the middleman processors would earmark the credit card deposits for items such as T-shirts, golf balls, even online flower delivery. That way, the banks profited from the transactions, Absolute Poker could still earn its rake, and the players could enjoy their poker.

In Brent's mind, even though the means were a little bit shady, when properly implemented, the result didn't seem to hurt anybody. The customers who were giving their credit card numbers to the processors were doing so willingly, because they wanted to play poker. And Brent didn't believe that the banks were actually being fooled; one day their credit cards were being used to deposit a million dollars in money earmarked for an online poker site; the next day, the same million dollars was deposited to purchase T-shirts and golf balls? They knew exactly what was going on, and they didn't care. It was just the way the business worked.

The banks made money. Absolute Poker made money. The players got to play poker. And Brent got an all-expenses-paid trip to Monte Carlo.

Brent came out of the bathroom and retrieved his tux jacket and bow tie from where they were hanging by the bedroom dressing table. Hilt looked ready to go; he was standing by the door to the balcony, bathed in the glow from the plaza lights, the reflected glory of the ancient casino. Hilt glanced at Brent, who had finally gotten his bow tie to sit straight and was straightening his sleeves over his cuffs, and smiled.

"You look like you're about to get married."

Brent blushed. He knew Hilt wasn't just talking about the tux, or the party they were about to attend. Brent had been opening up to Hilt during the whole trip from Costa Rica; Hilt knew that the new job wasn't the only thing that had changed in Brent's world over the past few months. Right before they'd left for Monte Carlo, his personal life had gone a little bit crazy.

He'd first gotten the news while in Vancouver—he'd gone to Canada as soon as Pete was there, and the team had gotten the new headquarters up and running, on the third floor of a sleek office tower in a prize corner of that city's financial district. Brent had been settling in for an indefinite stay in the Canadian city. The team had really set up a first-class operation, hiring a top-notch CFO with an MBA from one of the elite Canadian schools—a genial six-foot-three, 250-pound Fijian transplant who'd just ended a stint running the accounting department of a Fortune 500 banking giant—and Brent had been looking forward to living in a city where everyone spoke English and the power stayed on twenty-four hours a day. And then, quite accidentally, he'd run into a friend who'd just come from San José; the man had mentioned that he'd recently been at a party back in Costa Rica and had seen one of Brent's ex-girlfriends—a beautiful Colombian national Brent remembered fondly, even though they'd only been together a few months. The friend had casually mentioned that the woman had just had a baby boy—and then, still in passing, he'd said, "And he looks just like you."

It was something Brent couldn't ignore. An hour later he had called her up.

"I heard you had a baby," he'd said.

"I had your baby," she'd responded. "If you want, you can come see him. If you want to never call me again, that's fine too. It's up to you."

Brent had gotten on the next flight to San José. When he met the kid for the first time, he just sat there, staring at him, thinking he looked more like some sort of wrinkled alien than a baby. He'd actually poked the kid a few times, just to see if he was real. And then all he had wanted to do was hug him and hold him. Deep down, he knew. He was a dad, and nothing else in his life would matter as much as that.

Hilt might have been joking, but Brent knew, after the trip to Monte Carlo, that he wasn't going to be returning to Vancouver. He'd be going back to Costa Rica. And it was early to say for sure, but he didn't think marriage was out of the question. That kid had changed everything.

"Let's see how the party goes first," Brent responded, following Hilt toward the door. "If you treat me right, we can skip the wedding and go right to the honeymoon."

Five hours later Brent found himself stumbling down a cement spiral stairwell, out-of-his-mind drunk, searching desperately for a bathroom. In the portion of his brain that was still functioning, he knew he'd taken a wrong turn somewhere back in the casino, but the minute the fifth shot of sambuca had hit the back of his throat, all reason and sense of direction had gone out the door, and he'd become a man on a mission, navigating entirely on determined, if faulty, autopilot.

The party had been beyond extravagant: a buffet that seemed to go on for miles, offering everything from piles of stone-crab legs the length of baseball bats to vats of beluga caviar that could have filled a sandbox; four working bars staffed by a half dozen staggeringly beautiful bartenders, all amazonian Eastern Europeans who looked like they'd stepped off the set of a James Bond movie. Everyone in tuxes, and everyone on their best behavior. Except maybe Brent himself, who wasn't sure when he'd gone from happily buzzed to frighteningly blitzed.

When that last shot of sambuca hit, he'd known what he needed to do—just not where he could do it. He'd stumbled around looking for a bathroom for a few minutes, before pushing his way through a heavy curtain and into the stairwell—and now he was descending into what appeared to be the subterranean depths of the Monte Carlo Casino, looking for somewhere to throw up.

One more step, foot after foot—and suddenly, his leather loafer slipped on the concrete and he went down, landing on his ass. Then he was sliding forward, step to step, and the motion was simply too much for his stomach to handle. He leaned to the side and the warm vomit spewed out, splashing down the cement stairs right next to him.

Christ. He slid a few more steps down the stairwell, vomiting as he went—and then landed on the floor of a hallway, dazed, head spinning. There was a door right in front of him, another to his right. He knew he had to make a choice. One of them might be a bathroom, he supposed. Then again, it was just as likely that it was some sort of torture chamber, where they took people who had the gall to vomit all over the most famous casino on earth.

He slowly staggered to his feet. He was about to open the door directly in front of him when he heard the crackle of a radio from behind it; he couldn't understand the French words, but the radio definitely gave him pause. It sounded like the sort of thing a security guard would carry.

He quickly shifted right and headed for door number two.

It wasn't a bathroom—it was even better. A second later he found himself outside in an alley behind the casino. To his left, a narrow road curved up the hill back toward the plaza; elegant lampposts lit the way to his palatial hotel and the comfort of his marble bathroom and canopied bed.

Wiping vomit from his lips, he staggered upward, lamppost to lamppost. It dawned on him as he went that these were the same curves the Formula 1 cars would take, tires screaming against the pavement, as the entire city applauded and cheered.

In Brent's inebriated mind, he was suddenly one of those race cars, tearing through the city—it was a wonderful feeling, a high that felt like it was never going to end.

CHAPTER 24

AUGUST 2006

Pete read through the fax on his desk for the third time, but still the numbers didn't change. His gold-plated Cross pen—etched with the SAE insignia, a gift from the house when he'd handed over presidential duties to the president before Brent—danced in the air, rising and dipping precipitously between his first and second fingers; still, he couldn't bring himself to mark up the document. Because really, what was there to say?

The fax was only the first page of the report that the M&A fund had made of their nearly complete third-year financials, but it told Pete, and everyone else, everything he needed to know. The company's revenues had increased by more than 100 percent, year over year, and they were now funded at a valuation of more than one hundred million dollars. Since Pete was working in accounting and heading up an affiliate team, Scott had asked him to go through the report and add his input, point out things they might need to work on in the last few weeks before they were set to begin their IPO talks—but there was very little that Pete could think to add. One-hundred-million-dollar valuation, revenues exceeding forty million a year, and all of it increasing at a rapid pace.

Pete spun the pen between his fingers, its gold surface flashing in the sunlight that streamed in off the bay, reflected through the large picture window behind his desk. The view was spectacular, a straight shot through a corner of the financial district to the water, high enough to see over the low pincushion of office buildings, with sight lines to the mountains to the east as well. But in the past few months he'd gotten used to that view; the giant numbers on the fax were much harder to digest.

Still, he had to give Scott something. His salary had already been bumped up twice since they'd launched the Vancouver office and management had hired their new Fijian-Canadian CFO, who went by his first name, Shay. He began to scribble in the margins of the fax about dropping some of Brent's more disreputable payment processors when he heard a tapping on his office door.

He looked up to see Megann Cassidy, one of his marketing department heads—another recent hire, Canadian with an advanced business degree from McGill—standing in the doorway. Her face looked pale, and she was leaning into his office like she was trying to make herself even smaller than her five-foot-five frame.

"Mr. Barovich? I think there's something going on in the lobby."

Pete put his pen down on top of the fax.

"What do you mean?"

"Well, these two ladies came in a few minutes ago," she said, pausing as Pete nodded. Pete had seen them come in; he had been on his way back from lunch and had just gone through the glass doors that led from the lobby into the cubicled Marketing and PR Department when he caught sight of the two women. The truth was, they would have been hard to miss. Both were dressed in dark gray business pants, matching blouses, and heavy gray jackets. One was tall, at least five ten, the other around five seven. Both had dark hair—one in a tight ponytail, the other bunned up on top of her cubic head. Neither had been smiling, which was strange, considering it was Canada. Everyone seemed to smile in Canada.

"Well," Megann continued, "they just showed Lucinda their badges. They're agents with the RCMP."

Pete stared at her. RCMP: the Royal Canadian Mounted Police, Canada's version of the FBI.

"What do they want?"

"I don't know—" Megann started, her voice shaking, and then the phone on Pete's desk buzzed to life, signaling a call from the lobby. He waved Megann away, and she backed out of the office in full reverse, shutting his door behind her. Pete took a breath, then lifted the receiver.

"What's going on, Lucinda?"

The moment Pete heard Lucinda Palmer's voice, he knew that something big was happening. The secretary was as fierce as they came; in her midthirties, she favored leather skirts, high boots, and thick wool jackets, and she had been hired to man the front lobby specifically because she knew how to handle people without ever losing her cool. It was a lesson Pete had learned running numerous businesses since college: the customer may always be right, but he's often also an asshole.

But this was something different. These women weren't customers, they were federal agents. And Lucinda sounded like she was almost in tears. She told Pete that the women were asking for whoever was in charge. She'd tried to connect them to Shay, but Shay wasn't answering his phone. Shay's secretary had told her that he wasn't there, but Pete knew that wasn't true—Shay had returned from lunch a few minutes before he himself had. So if Shay was avoiding them—well, maybe that wasn't such a bad idea.

"Tell them I'm sorry too, I'm just on my way out."

He quickly hung up and rose from his desk. He crossed to the door of his office, quietly pulled it open, took one step out into the hall—and saw his newly hired CFO about five yards away, down on his hands and knees, literally crawling behind the row of cubicles. The heavyset Fijian cut an absurd figure; all six foot three of him, dressed in a tightly fitting tailored suit and tie, his face inches from the carpet as he tried to stay below the sight line of the women on the other side of the glass wall that separated the cubicles from the lobby.

Pete realized with a start that Shay was heading for the fire exit. Within minutes, the CFO crawled the last few feet to the fire door, then raised himself to a crouch. He looked back at Pete and made a frantic gesture with a tanned hand, telling him to follow. Then the man put his weight into one thick shoulder and leaned into the door. The door came open, and then Shay was gone, sprinting heavily down the fire stairs.

Pete swallowed. He glanced toward the glass lobby wall; he could barely make out the two federal agents, the ponytail and the bun, as they berated Lucinda. He knew he had to make an immediate decision, because any second now, those women were going to push past the secretary—and God only knew what might happen next.

So he did the only thing he could think of, and dropped to the floor. Hands in front of knees, he crawled toward the fire exit as fast as he could.

Five minutes later he was standing next to Shay at the window of the Starbucks across the street, watching the RCMP agents rifling through his office. He had no idea what they were looking for, or even if they had the legal right to do what they were doing. He and Shay had put three calls through to Scott and the Absolute Poker legal team, but they hadn't heard back yet. He was pretty sure Scott was actually on vacation somewhere in Europe with a girl he'd begun seeing seriously; Pete had heard from Garin that the girl was a real head case, and not a long-term potential—par for the course, since Scott's relationships usually involved so much drama. At the moment, Pete wouldn't have given a damn if she were a spear-toting harpy. He needed to talk to Scott or their lawyers, because he had no idea what they were supposed to do next.

Another twenty minutes went by, he and Shay standing in the Starbucks in near silence, and finally the two RCMP agents left the way they had come, through the building's front doors. Pete and his CFO waited another ten minutes, then decided it was safe to go back inside—via the elevator this time, not the fire escape.

When the elevator deposited them back on the third floor, a loud buzzing sound greeted them. It was coming from the financial wing, where Brent had set up his processing department before he'd gone back to Costa Rica. It took Pete a full minute to realize what the buzzing was—and when he did, his face went a shade paler. *Paper shredders, running nonstop.* He had no idea what, exactly, was being shredded, or by whom. And he decided that he didn't want to know.

Instead, he led Shay into the lobby, where they found Lucinda slumped behind her desk, in tears. After she'd gone through a box of tissues, she told Pete that the RCMP agents had first demanded to know who was in charge—and then they'd demanded to know what, exactly, was going on in the third-floor offices. Lucinda had been both terrified and bewildered; the RCMP couldn't be there to arrest anyone, because they weren't even sure what sort of business the company was in. In fact, as far as Pete could gather from the questions they'd asked Lucinda, the agents were only there because a high-level Canadian official was convinced something illegal had to be transpiring, simply because of the amount of money that was moving through the building in such a short period.

It was ridiculous, sheer harassment—and yet Pete could understand the terror Lucinda was feeling. When she told him that fully three-quarters of the staff had already quit—dropping their ID cards on the lobby desk on their way out—he realized that this wasn't the sort of trouble they could simply get past. He looked at Shay and could see that his CFO was thinking the same thing. If the RCMP had sent agents to harass them in the middle of the day, there was a good chance that it was just the beginning.

Soon there would be more agents, search warrants, maybe even confiscations.

Pete also knew that in the back of those offices there were computer servers engaged in running their hundred-million-dollar business; if the RCMP confiscated those servers, even under illegitimate orders, it could destroy Absolute Poker.

He thanked Lucinda, did his best to comfort her, then accepted her resignation. He told her that they would pay everyone who had quit double their expected severance. Then he pulled Shay aside and lowered his voice. "We need to find a U-Haul truck—fast."

Ten minutes past midnight, and Pete rolled the oversize U-Haul the last few yards into the parking lot behind his office building with the engine off. Shay, in the seat next to him, hunched as low as a 250-pound person could hunch, was sweating beneath his three-piece suit.

"Are we really going to do this?" Shay asked, but Pete didn't answer. He had already spoken to Scott twice, and he knew that they didn't have a choice. As wrongheaded as the RCMP seemed to him, there was no reason to believe that they wouldn't go after their servers next—and that simply couldn't happen. As far as Pete knew, the company hadn't broken any Canadian laws, and the servers were their property. The computers themselves were worth six figures, but the information on them was worth potentially many millions.

Pete got out of the truck, Shay behind him. They made their way through the darkened lobby. When they got to the elevators,

Pete swiped his key card into the after-hours security slot—and instantly discovered that his key card no longer worked. Another really bad sign. He looked at Shay, wondering how they were going to get through, when he heard noise from the far corner of the lobby. He turned to see a cleaning man in a blue janitorial uniform dragging a heavy vacuum cleaner toward the thick shag carpet.

"Wait here," Pete said to Shay. Then he hurried over to the janitor. The man recognized him, of course; Pete and Shay had often worked late since setting up the new offices. It wasn't unusual to see them there well after midnight.

"I think my card got demagnetized," Pete said, offering as easy a smile as he could. "I wonder if you could help me out."

Before the man could answer, Pete pulled out his wallet and took out a fifty-dollar bill. The janitor looked at the money, then shrugged, smiling back.

"Happens all the time," he said. And he handed Pete his own key card. "Just bring it back on your way out. I should be here all night."

Arriving on the third floor, Pete and Shay went to work like men possessed. They found a pair of huge metal roller carts in a supply closet behind Lucinda's desk, then went through the offices, grabbing everything they could get their hands on—not just the huge servers, which were boxy steel cabinets the size of small tables, but also laptop computers, monitors, phones, and any notebooks that looked important. All of it needed to be done in one trip—two people, two roller carts, and enough stuff to fill an entire office floor. Since they could fit only the crucial material, they had to leave behind hundreds of thousands of dollars' worth of equipment. But there was no choice. An hour later, having gotten as much onto the carts as they could manage, they headed for the elevator. Then Pete remembered one more thing. He rushed to his office and retrieved his gold SAE pen. Shoving it into his jacket pocket, he ran back to Shay, and they took the elevator back down to the lobby.

Adrenaline pumping, they began to push the two roller carts across the freshly vacuumed thick shag rug. Just as they reached the doors leading out to the parking lot, the janitor rushed toward them. Pete remembered the key card, yanking it from his pocket and offering it to the man. The man grabbed it, then lowered his voice.

"I just got a call from the owner. He said the cops were on the lookout for anyone who worked in your company. I told him you just left—so you've got about five minutes at the most to get out of here. Good luck."

Pete and Shay sprinted the last few yards to the U-Haul truck, pushing the metal carts in front of them. They threw the computers and equipment into the back, trashing most of it in their haste, then slammed the doors and started the engine.

They could see police lights flashing in the rearview mirror as they pulled away from the office building.

It was 5 A.M. by the time they finally got Scott on the phone. They were in Shay's apartment, in a corner of his brightly lit kitchen, standing a few feet from the refrigerator—because that was where Shay's only landline was connected and neither of them wanted to use their cells.

As Pete had guessed, Scott was somewhere outside of Paris; there was a woman's voice in the background, but it was hard to tell from her accent where she was from. Pete had barely gotten three words across to his friend before Shay had calmly taken the phone out of his hand. Leaning back against a granite kitchen counter, the oversize CFO methodically told Scott exactly what had happened. He explained that after they had left, they had made some inquiries and had found out that the RCMP had indeed opened up an investigation.

Pete couldn't hear exactly how Scott responded, but obviously he wasn't taking the situation quite as seriously as Shay wanted—because one second later Shay's face turned bright red, and everything calm and methodical about him went right out the fucking window. He began screaming into the receiver, "*You better fucking figure this out, and get us legal right now! You better get us some fucking lawyers and get this shit under control—*"

Pete wrestled the phone out of Shay's hand. He gave the man a moment to calm himself down, then put the receiver to his own ear and listened to Scott, who was obviously trying his best to remain rational in the wake of their CFO's meltdown.

"What do you want us to do?" Pete finally asked. Shay had slumped to the floor, shaking his head.

"I want you to get your asses back to Costa Rica," Scott responded quietly. Then he hung up the phone.

Pete replaced the receiver, then dropped to the kitchen floor next to Shay and rested his chin in his hands. Shay glanced at him.

"Are we doing something illegal?" Shay asked.

Pete didn't answer. He didn't know what to say. He had sold his whole life and moved his wife and dogs to Vancouver, he had been there a few months, and the whole thing felt like it was blowing up around him. He was only twenty-nine years old, but even at that young age he knew that this wasn't the way a billion-dollar company was supposed to be run. And up until that moment—that vivid, seemingly frozen moment with his CFO on his hands and knees and federal officers standing in the lobby—Pete had thought that Scott's company was just like any other rapidly expanding Internet corporation, providing a service for millions of people, revenues skyrocketing as the world moved online. In just a few weeks they were supposed to start preparing for a billion-dollar IPO. Hell, there was still a fax sitting on his desk in his ravaged office with numbers telling him that the company Scott had founded was worth more than a hundred million dollars at that very moment, with revenues in the tens of millions.

But now he had to wonder.

What, exactly, had he gotten himself into?

And where did they go from here?

CHAPTER 25

Two weeks later, on a Friday night in September, Pete still hadn't quite gotten a handle on what happened in Vancouver. The next few days had been a whirlwind, setting up the rescued servers in an apartment near their raided offices so that the accounts could keep spinning—a hundred-million-dollar company basically run out of a one-bedroom, four-hundred-bucks-a-month walk-up—and then hightailing it out of Canada, wife and dogs in tow. As far as Pete knew, he personally wasn't in any trouble. He hadn't broken any laws that he knew of, and there hadn't been any Canadian federal claims made against Absolute Poker either. But Scott hadn't wanted to take any chances—if the Canadians didn't want them there, they would leave.

That's how it had been with Absolute Poker from day one: if someone wanted to regulate them, they were happy to comply with any and all rules. In the UK, online poker was indeed taxed and regulated, and as Party Poker had proven, you could successfully IPO on their exchanges. In many other countries as well, Absolute Poker was licensed and approved, paying whatever fees were necessary—or, in other cases, simply blocking people from playing when a country made it clear that it wasn't legal to play from there. Players from China, for instance, weren't accepted by Absolute Poker; it was as easy as blocking any computer with a Chinese ISP.

But Canada, obviously, was momentarily unclear about how it viewed the business—so for now, Vancouver was done. Pete was in Costa Rica, winding down the Vancouver operations, trying to help the team ready a first-world company for a billion-dollar IPO in a third-world setting. It wasn't easy, but he had no reason to believe they couldn't make it work. Everyone they talked to, from the bankers to the lawyers, believed that the company was weeks away from being valued at over a billion.

He glanced at his watch and realized it was nearly nine thirty; Brandi was going to kill him if he stayed any later. She didn't like living in San José, and she had made it abundantly clear that the stay was going to be temporary. Pete had assured her that as soon as that IPO went through, he would have enough shares in the company for them to live anywhere in the world she wanted.

He flicked off his computer and rose from his chair, his eye-line rising above the edge of his cubicle. Scott had given him one of the larger prefab squares; he didn't have real walls or a window, but at least he had two computer screens and easy access to the kitchen, which was really a pair of refrigerators and a sink out of which spurted water he wouldn't have washed his dogs in.

He reached for his briefcase, tucking a few closing reports on Vancouver into the side pocket, and was about to head toward the stairwell that led down to the parking lot when he noticed a dull yellow light coming from the far end of the office. Someone was in the very last cubicle, hunched over a computer screen.

Pete knew Scott and Hilt had headed out earlier, Shane was at home, and Garin had taken a three-day weekend to visit relatives back in the States. It could have been one of the Costa Rican employees, but he doubted it; it was extremely unlikely that any of them would have been in the office on a Friday night, especially since it wasn't raining outside—a brief gap in the unrelenting rainy season—and most would be on their way to the beach.

That left Brent. Which was odd as well; since returning to Costa Rica to start his new family, Brent had settled into at least a semblance of a more normal working life—a Friday night was an unusual time to find him at a computer screen.

But as Pete circled around the cubicles and approached the glow, he saw that it was indeed the youngest member of their team. Brent didn't look up as Pete slid into the cubicle behind him, and Pete was about to give him a little shove to wake him up when he saw the way Brent was gripping the edge of the desk beneath his computer's keyboard.

"Hey, man, you okay?" he asked. Brent didn't glance up from the screen. Pete looked toward the glow and saw that the screen was filled with video—an assembly hall of some sort, with wooden chairs and benches, most of them empty, and a scrawl of writing beneath. Pete read some of the words—and saw immediately that Brent was watching a feed, either recorded or live, from C-SPAN. Pete had never watched C-SPAN before, but he could tell that what Brent was looking at was some sort of U.S. congressional session.

"I don't think that I am," Brent finally answered. "I don't think any of us are." His voice sounded strange.

Pete put a hand on his shoulder. "What are you watching? Did something just happen?"

"As far as I can tell, they just voted through something called the Unlawful Internet Gambling and Enforcement Act of 2006. I've gotten like twenty e-mails about it in the past ten minutes. The UIGEA, they call it."

Pete squinted at the screen, reading more of the scrawl beneath the picture.

"It says they just voted on a Safe Port Act. It says it's a bill about protecting our ports from terrorism," he said.

"Yeah, that's the thing. They tacked the antigambling bill onto the Safe Port Act, because they knew nobody would vote against an antiterrorism bill. Most of the people who voted it through didn't even read the antigambling addendum—they just voted to protect our ports. The bill itself, it's basically the same antigambling bullshit a couple of moralizing senators have been failing to get anybody to take seriously for years, but now it's passed through Congress. On the last day of the congressional session, with almost nobody around to even take a look at what they're voting on."

Brent hit keys on his keyboard, pulling up a handful of e-mails he'd gotten from friends across the industry who had been monitoring the legal back-and-forth that had led up to the congressional vote. Pete speed-read through the summaries—and as he got to the gist of each e-mail, he felt the color leaching from his face.

According to Brent's experts, the UIGEA was really the

brainchild of two conservative senators—Bill Frist, a Republican from Tennessee, and Jon Kyl, a Republican from Arizona—who'd come up with the ingenious plan of attaching it as a last-minute amendment to the Safe Port Act—no matter that Internet gambling had nothing to do with protecting U.S. ports from terrorists. The two antigambling senators, who had run for their positions on morality platforms, knew that trying to take down a pastime that millions of Americans were already enjoying was too difficult, so they'd concocted what was essentially a sneak attack.

The bill didn't make *playing* online poker illegal. In fact, the bill didn't even make running or owning an online poker site illegal. What the bill made illegal was dealing in online gambling proceeds. Specifically: "*The Act prohibits gambling businesses from knowingly accepting payments in connection with the participation of another person in a bet or wager that involves the use of the Internet and that is unlawful under any federal or state law.*"

It wasn't entirely clear from the bill what constituted a "gambling business," but the addition of the phrase "unlawful under any federal or state law" seemed to give a lot of leeway to prosecutors who wanted to go after an Internet company. All you'd need to do would be to find a state with a particularly harsh definition of *gambling*—and you could use that definition to go after any website that took money from a player living there.

Poker, Pete firmly believed, was for the most part a game of skill. But he also knew that in a few states—New York, for example—there were some pretty wide definitions of what constituted illegal gambling, definitions that could seem to include even games that were mostly skill. So what, exactly, did this mean? It was a bill; it had now gone through Congress. It still had to be signed by the president, which would take some time. But what did this *mean*?

"This seems crazy," Brent was saying, tapping the screen. "It doesn't target the players. It doesn't even target the game. But it makes the movement of money illegal, if that money is used for illegal gambling. But how do you define illegal gambling? Every lawyer we've talked to says under the Wire Act of 1961 it doesn't mean poker. But for players playing from the U.S.—I guess this is saying it varies state by state?"

Pete rubbed a hand through his hair. "Have you talked to Scott?"

"A few times since these e-mails started coming in. He's pretty pissed about this too. I think some of the shareholders are running around like the building just got set on fire. The lawyers, though—most still believe the legislation isn't clear, especially for a private, international company like ours. And on top of that, the lawyers have brought him and Hilt a fairly radical offer, which they're kicking around."

"What's that?"

"You ever heard of Joseph Norton—the Canadian Indian chief?"

Pete raised his eyebrows. It was a name out of left field, but having just dealt with the Royal Canadian Mounties and taken a look at Canadian gambling law, he had indeed heard of Joseph Tokwiro Norton—the former grand chief of the Mohawk Council of Kahnawake, near Montreal. Norton was a bit of a legend in Canada. Story was, he'd once stopped an RCMP raid by literally blocking federal troops from crossing a bridge into Mohawk land. An actual armed standoff—First Nations against federal agents with guns and tanks. Norton had negotiated a peaceful resolution and had become a hero in his community. In recent years he had become a successful businessman and had established the Kahnawake Gaming Commission, which licensed and regulated Internet gaming sites. Because he operated out of designated First Nations territory, he wasn't bound by Canadian law; even if there had been any clear rules against poker, they wouldn't affect his business.

"What about him?"

"I think he's about to become the owner of Absolute Poker. He's approached our legal team, made it clear he's interested in buying the company."

Pete looked at Brent. "We're going to sell the company to an Indian chief?"

But when he thought it through, it made sense. Certainly, it added a step between them and this UIGEA bill, and there was no question that under Mohawk law an online poker company would be legal. However, from what Pete was reading, that wasn't exactly what this bill was going after. On a straight reading of the UIGEA, running an online poker company wasn't illegal. Playing online poker wasn't illegal. The only thing that was illegal was accepting money from U.S. players. They were going after the flow of money—and the more Pete thought about it, the more he realized that this was going to change everything.

"It's like Prohibition all over again," he said. "Every company is going to have to go through the same thought process—do we shut down now, just close up shop? Do we cut out of the U.S. market and only go international—even though that would just about kill us, or any of our competitors, because as an industry, eighty percent of our players are American? Or do we just keep doing what we're doing, take our chances, and hope this is more smoke than fire?"

The way Brent was still gripping the desk beneath his computer, Pete could tell they were sharing the same thought. They were weeks away from starting the process of their IPO, and this was going to ruin everything. This bill, this UIGEA, snuck through on the tail of some antiterrorist Safe Port Act, wasn't just smoke, and it wasn't just fire.

It was Armageddon.

CHAPTER 26

The panic had reached epidemic proportions; Scott could see it in the eyes of every person he passed as he strolled the last few yards through the huge, warehouse-like conference center's main floor, toward the elevators that led into the hotel. They all looked as though they were melting down inside, succumbing to a fever so fierce and virulent, it threatened to consume them.

Only when the elevator doors shut behind him, leaving him alone in the brightly lit, thickly carpeted steel box, buffeted by invisible waves of Spanish Muzac—some version of a Beatles song, Paul McCartney run through a vocal translator and backed up by what sounded like a mariachi band—did he let his body show the toll the last two weeks had taken, allow the true weight he'd been carrying around since the bill had gone through Congress to settle across his taut features.

It had been an emotional roller coaster, from the first minute his phones had lit up with the news about the UIGEA sneak attack—in Scott's mind, the act had come out of nowhere, tacked onto that Safe Port Act like a legal version of Pearl Harbor, with no warning to anyone in the industry—to the day he and his team had left for Barcelona, for the Internet gaming industry's annual conference. They had planned the trip months earlier, as had everyone else; but nobody could have foreseen that the conference—usually an event marked by partying to excess, good food, and plenty of optimism in an industry that was minting money day by day—would become ground zero for a community facing what appeared to be a mortal blow.

Scott coughed, trying to clear his throat as he watched the digital display on the elevator count upward. He had been talking nonstop since the bill had gone through, and he was in real danger of losing his voice. At first, the calls had all been the same—terror-stricken associates, competitors, and shareholders telling him that it was all over, that the gaming industry, as he knew it, was finished. Even Scott had joined in on the talk of gloom and doom; he'd immediately called Phil, his dad, and told him that it felt like they were done, that they'd have to shut down operations and just move on.

But by the end of the weekend, the calls had started to change in tenor, especially the ones coming from the many lawyers his company employed, who seemed to be split down the middle on what the UIGEA's passage really meant. To roughly 50 percent of the lawyers he spoke to, as a private, international company centered on a game primarily of skill, it wasn't clear that they were within the UIGEA's jurisdiction. Sure, the safe thing to do would be to shut down U.S. operations—which, for Absolute Poker, with most of its players coming from the States, would essentially mean closing up shop—but if they continued to operate, what could really happen? Would Scott, personally, be breaking the law? Not Costa Rican law, for sure. Not international law. Would he be breaking U.S. law?

It wasn't entirely clear. And when, at the urging of shareholders, he and Hilt came up with the plan to sell the company to Norton and the Kahnawake Gaming Commission—for a $250 million promissory note and a portion of future earnings—it seemed a responsible and legitimate way to separate them further from that jurisdiction; certainly they weren't breaking any Canadian First Nations tribal laws by running an online poker company.

It was tough, mentally—the idea of selling the company, parting, even in a small way, with the entity he had built from an idea into a worldwide brand. But there was no denying that the world he lived in was now in a state of flux and chaos. One minute he was weeks away from being a near-billionaire; the next, he was facing a confusing legal battle that nobody he talked to could even give him a straight answer about.

Whatever he felt about the future of Absolute Poker, there was no question that the industry as a whole was reeling. The minute the bill passed, shock waves ripped through the online poker world. Party Poker, which after its multibillion-dollar IPO had become the biggest site, cut and ran from the American market. It paid a $1.05 million fine to the U.S. government and agreed to exit the U.S. market in exchange for amnesty from any future legal actions. And as a result, its value had gone from around $12 billion to around $2 billion—overnight.

The other publicly traded companies followed suit, and really, they had no choice, because they had to answer to public shareholders who wouldn't face the risk of legal action, no matter how arbitrary it seemed from Scott's point of view. In the blink of an eye, dozens of companies shut down U.S. operations, basically giving up billions of dollars in revenue at the stroke of a conservative Republican senator's pen.

Although Scott and Hilt initially discussed doing the same, they eventually decided not to pull the trigger—at least not yet. For one thing, the venture capitalists and bankers they had been working with to capitalize themselves in preparation for their IPO basically forbade them from exiting the U.S. market. At that point the company had about four hundred employees, and with the vast majority of its revenues coming from the United States, it would be insolvent if it left. Further, UIGEA's jurisdiction arguably wasn't any more clearly damning to Absolute Poker than the original Wire Act had been—even the WTO was in the process of declaring online poker exempt from gaming statutes in a suit brought by Antigua against the United States for hampering its island-based gaming businesses. It didn't seem right that they should commit financial suicide based on what they saw as shaky law.

But even so, Scott knew that post-UIGEA, nothing would ever be the same. The way he understood it, his board of directors had been taking feedback from the lawyers, and had voted that the company was to maintain its U.S. presence. Scott felt left with no choice but to uphold that vote—after which, even though they seemed to have gotten what they wanted, 90 percent of the board had resigned, including Shane, along with a large percentage of the U.S. citizens who worked for the company.

Shane had decided to step down from his position—from the board, and from Absolute Poker as well—not because he believed the UIGEA prohibited Absolute Poker from doing business with U.S. players, but because he simply wasn't willing to risk facing prosecution or never being able to return to the United States.

Nobody wanted to be a cowboy, flaunting the law, even if that law seemed unfair and unclear. Scott felt trapped by the board's decision, but he also knew that if they did somehow go forward, there would have to be rules in place for the U.S. citizens who remained at the company. For the foreseeable future, lawyers told them that they'd probably have to restrict U.S. travel, maybe even ban it altogether. If prosecutors in the United States were going to try to go after them, they'd be looking for Americans on the inside of the poker companies to help build those cases.

Scott clenched his jaw as the elevator slowed near his floor. That he even had to think this way, like he was some sort of criminal for running an Internet poker company—it just seemed so unfair. Why was this happening? Why pass a bill like this, essentially trying to turn a moral issue into a legal one, rather than regulate them, like with cigarettes and alcohol, or give them a chance to regulate themselves? And why didn't anybody see how hypocritical this was? Nearly every state in the United States ran lotteries, which were clearly gambling, and nobody had a problem with that. Many states allowed horse racing, which was clearly gambling, and nobody had a problem with that. Vegas was built on casinos—hell, they'd rallied around the UIGEA, because it protected their established gambling interests and cut down their competition—and nobody had a problem with that. Poker itself was legal in most states, and nobody had a problem with that.

Christ, Scott could go on E*Trade that afternoon, open an account, and roll his entire 401(k) into it, start gambling on the stock market with no advisor and lose everything—and nobody would have a problem with that.

But online poker? Bill Frist was going to rally the American legal system to protect college kids from playing poker over the Internet? To Scott, it was the height of hypocrisy.

The elevator doors whooshed open, and Scott nearly tumbled out into a narrow hallway, with thick carpeting and sconces on the walls. He caught sight of Hilt immediately, at the entrance to the small, windowless conference room, standing with six other people; most looked like lawyers, in suits and ties, carrying briefcases—but right next to Hilt was Greg Pierson, the head and founder of Ultimate Bet.

At the sight of Hilt and Greg, Scott felt his shoulders rise, if only an inch or two—because even in the midst of all the chaos, terror, and misery of the moment, there might just be a silver lining.

Back when Scott was struggling to find a way to get Absolute Poker to the next level, he'd called on Greg for advice; now, with the industry imploding and the ground reeling beneath their feet, Greg had contacted him.

Ultimate Bet, like Party Poker, was a public company. By then, it had grown to be one of the biggest in the business—but like Absolute Poker, most of its customers lived in the States. If it shut down that part of its operation, Ultimate Bet wouldn't survive. Greg was in a real bind; he couldn't ignore the UIGEA, but if he responded the way Party Poker and the rest of the public companies had, it would be corporate suicide.

So he'd turned to Scott, just as Scott had once turned to him.

And over the past few days, Scott had come up with a solution. A crazy, wild, impossible solution. It wasn't going to undo the dramatically destructive results of the UIGEA passage, nor would it erase the threat of what might come because of it—but it was a response that would take advantage of the situation in the best way possible.

Scott painted a calm and cool mask across his features as he hurried down the hallway and followed Hilt, Greg, and the group of suits into the hotel conference room.

Thirty hours later, with $130 million changing hands, a deal was struck. A deal that instantly changed the landscape of an entire industry and put Scott and his friends in the dead center of a brand-new—and exceedingly dangerous—world.

It was sometime after 2 A.M., but it was impossible to tell for sure, because there weren't any windows. Hell, from where Brent was sitting—sunk waist-deep into a huge beanbag chair in the center of a circular pit of overstuffed pillows, thick velvet drapes, and wicker tables surrounded by ornate, Arabian-looking hookahs bubbling out mysterious blue-smoke concoctions—the place didn't seem to even have walls. The ceiling curved upward, mosque-style, and was dripping with colorful strips of fabric that waved in an artificial desert breeze provided by huge air-conditioning vents. The club was vaguely Middle Eastern, but Brent couldn't be sure, because he wasn't entirely certain where he was. He didn't know Barcelona at all, and the streets of the damn city were as circuitous as those of downtown San José.

He'd been led to the place from the hotel by the three young men who were sharing the hookah pit with him; two were his counterparts from Ultimate Bet, about his age and dressed in similar club garb—black shirt, dark jeans, and a plastic wristband from the mixer back at the conference hotel—and the third was from PokerStars, which, now that Party Poker had dropped out of the American market, was the biggest company on the block. All three did the same job at their respective companies; like Brent, they were in charge of payment processing—which meant that all of them were about to see their worlds turned upside down.

In this new landscape, payment processing was going to change—in a big, big way. Although Neteller claimed it was going to stay in the business—UIGEA seemed unfair and unprosecutable, it said—Brent and his counterparts were very likely going to have to find many new middlemen, and soon, if they were to continue to collect deposits from American players. This meant dealing with the shadier and shadier outfits that would spring up to fill the void left by those who fled fearing criminal prosecution—no doubt charging exorbitant fees and turning what was already a shaky part of the business into something much dirtier. Because anyone willing to handle the flow of money in and out of Internet companies after UIGEA was risking being labeled criminal—and anyone willing to take that risk was probably already a little dirty to begin with.

Brent didn't want to think about what that made him; his name was on most of the financial transactions that had gone through his company since he'd been handed that stack of paper giving him the financial processing job. If anyone at Absolute Poker was flouting the new bill, it was him.

At the moment he wanted to think about anything other than that. He took a deep breath, letting the thick, flowery scent of Moroccan incense blend with whatever the hell was bubbling out of the hookahs, and was reaching for what had to be his fifth shot of rum when a shadow leaped across their pit of beanbags and another guy joined their group. He was lanky, and wearing a poorly fitted suit. *He must be a salesman of some sort,* Brent thought, and as he shook hands around the hookah, Brent heard him introduce himself as another Ultimate Bet employee, in the Marketing Department. His name was Joe or John, but Brent had never been great with names, especially when his throat was full of expensive Spanish rum.

He was barely paying attention as Joe or John—or maybe Joel—starting chatting excitedly about the conference, about some big development that had just been announced on the exhibit floor. He barely noticed as the other Ultimate Bet employees seemed to stiffen, then chatter loudly among themselves, their faces a mix of awe, excitement, and a little fear. He barely noticed anything—until he realized that everyone in the pit was now looking at him, and then he finally caught a few words that were still hanging in the thickly scented air.

"Um, what was that?"

"Did you hear the news?" the new guy said, his voice high-pitched. "Ultimate Bet just got bought out."

"Wow," Brent said, unfazed, reaching for another shot. "By who?"

"By who?" the guy responded. "By you."

Brent stopped, his hand frozen above the shot glass. "No way."

One of the other Ultimate Bet guys nodded. "It's true. I just checked my phone, got three texts from our CEO. Guess what, Brent—now we all work for you."

Brent's head was still spinning as he sat in the waiting area of the Barcelona airport, just inside Security. He was hungover as hell, maybe still a little drunk, and he smelled of alcohol, hookah smoke, and decidedly citric perfume. The perfume was French, though he was reasonably certain the girl had been Spanish; he was also pretty sure she was a waitress from the Middle Eastern club, though the night had gotten very hazy after the hookah pit. He'd woken up in a tiny first-floor apartment in the El Raval section of the city—which, he knew from a guidebook, was essentially the red-light district, a bit seedy, but also home to Barcelona's artistic community. So maybe the girl in the leather skirt, leather boots, and nothing else on the futon next to him was an artist? He hadn't woken her up to find out for sure. He'd just checked his phone, seen the twenty or so texts and calls from Hilt telling him he was going to miss his flight if he didn't get to the airport right away, and yanked on his dark jeans and shirt and gone in search of a taxi.

And somehow, he'd actually beaten Hilt to the airport—because he was already sitting there in the waiting area by his gate, his little carry-on bag under his chair, when he saw his brother's main business partner come out through Security and hurry toward him.

"We've still got a few minutes," Brent croaked, his throat still feeling the effects of all that rum. "The flight got delayed. They say ten more minutes."

Hilt nodded, then dropped into the empty seat next to Brent. Hilt looked even worse than Brent felt; it was obvious he hadn't slept at all. From what Brent had heard about the negotiations with Ultimate Bet, it wasn't surprising. Normally, a merger like that would take months of due diligence, haggling, give-and-take. Hilt and Scott had gotten the deal done in thirty hours. A promissory note of around $130 million, $5 million in cash, a profit-sharing agreement—and just like that, Absolute Poker/Ultimate Bet had become the third-largest online poker site in the world. Alongside PokerStars and Full Tilt Poker, the two other major private companies that had decided to stay in the U.S. market post-UIGEA, Absolute Poker was suddenly going to dominate a landscape that had been left open by the desertion of Party Poker and the rest of the publicly traded companies.

Which meant that literally overnight, Absolute Poker's revenues were going to go from around $200 million a year to as much as five times that amount. The amount Brent, personally, would facilitate the processing of would go from $25 million a month to closer to $75 million. It seemed like an insane, impossible number. And not only that: he was going to have to do so in an environment that was going to be more and more like the Wild West, because the rules were now completely uncertain and the middleman processors were going to get even shadier.

"You're going to have to make a decision," Hilt said, continuing a conversation he'd been having with Brent's voice mail. "And if you do decide to stick with us, you'll have to follow some new rules. First, we have to ask that you not go back to the U.S.—it's just too confusing, and we don't know what will happen. So nobody with an American passport is allowed to go through the

U.S. anymore, at least for the foreseeable future. We believe that in the long run, this bill won't hold up—and even if it does, it won't be prosecutable. But for now, if you go overseas, it has to be a direct flight."

Brent nodded. It was a strange thought, not being able to return to see family. It was giving up a lot. Weddings, funerals, birthdays—he'd be cut off from the people he'd grown up with, from parents and uncles and friends. For all of them, it would be an intense sacrifice. He knew that a good portion of the company's American employees had resigned right away, but Brent didn't want to walk, at least not yet. The thing was, his hands were already a little dirty; he'd been running payment processing for some time, and he knew that many of the little things his middlemen had done to get through the banks' sometimes arbitrary rules were probably technically illegal. Forms that earmarked deposits for the purchase of golf balls and T-shirts, instead of gaming, were going to look pretty bad, even if they hadn't hurt anyone, even if the banks had known what was going on.

"If you decide to leave," Hilt continued, "nobody is going to have any problem with that. But if you stay, we're going to bump your salary to twenty-five thousand a month. We're going to jump from four hundred employees to close to a thousand. And now we have offices in Costa Rica, Toronto, Vancouver, Malta, Montreal, Korea, the UK—the list is endless. You're going to be responsible for many of them. As well as almost two million dollars a day in player money."

Brent's hangover seemed to disintegrate as he listened to the numbers. *Twenty-five thousand dollars a month.* It seemed like a fortune. If things stayed like that for a while, he was going to get rich. *Two million dollars a day.* Hell, all of them were going to get rich. He had no idea how much Scott and Hilt were getting paid. He couldn't begin to imagine. There was no more board, and a Canadian First Nations tribe was officially running the show. He wasn't even sure what roles Scott and Hilt were going to play going forward; for all he knew, they were going to fade into the background and let the company run itself. It was a true empire now, with huge revenues—and a worldwide structure.

"I'm in," Brent said. "One hundred percent."

Before Hilt could shake his hand, a voice broke over the airport intercom in Spanish, then in heavily accented English. Their flight to Malta was about to start boarding. *Malta.* Brent was certain he couldn't find the place on a map—and yet now he was on his way to visit a payment office there—an office that he now ran. A day ago he had been booked on a flight back to Costa Rica, toward his cubicle and his Neteller accounts and his handful of little headaches. Now he was on his way to what Hilt warned him could extend into a three-week around-the-world trip—to offices in a half dozen countries, offices that Brent now ran.

Brent rose to his feet, excited and scared. He reached into the front pocket of his jeans, looking for his passport. Instead, his fingers touched something that felt like crumpled plastic wrap. He pulled it out, looked at the thing—and then his entire body froze.

In his palm was an eight ball of cocaine—three and a half grams of the white powder, rolled into a little plastic-wrapped ball.

"Damn," Hilt whispered, staring at him.

Brent clenched his hand closed, then looked around, at the airport waiting area. Travelers were lining up at the gate not ten feet away. In the other direction, other travelers were still spilling out of the security area, some repacking their bags and putting their shoes back on.

Brent wasn't sure how the coke had gotten into his pocket, or even whose it was. Maybe the Spanish girl with French perfume, maybe one of the Ultimate Bet guys, God only knew. And he had absolutely no idea how he'd gotten it through security.

He forced his legs to start moving and quickly crossed to a nearby garbage can. Then he tossed the little plastic bag inside and rushed back to where Hilt was standing, still staring at him.

"You've got to be the worst Mormon I've ever met," Hilt said, bewildered.

Brent grinned back at him. Together they joined the row of travelers lining up for the ninety-minute flight to Malta.

CHAPTER 27

"Completely insane, man." Brent's voice croaked out of the speaker on the polished mahogany desk in Scott's home office, mingling with the gurgle from the marble fountains out on the manicured front lawn drifting in through the open French doors that led out onto the two-hundred-square-foot balcony. "Malta to Dubai, then Singapore, Hong Kong, and all the way to Frankfurt. It's like the Gumball rally in reverse."

Scott laughed. He was seated behind the desk, his white collared shirt open to the third button, a Corona in his left hand, his legs up on the desk, flip-flops on his feet next to the telephone speaker. In his lap was a .38 caliber Smith & Wesson revolver. As he listened to his brother his right hand rested lightly against the gun's grip, his finger on the safety. The cool touch of the metal felt strangely comforting against his skin.

"And every meeting has been just like Malta, back in the beginning," Brent continued. "We get in there—this building had to be like three hundred years old—and everyone just stares at us. 'What do we do?' they ask. 'What do you normally do?' we ask back. 'Just keep doing whatever that is, only now you're doing it for us instead of UB.'"

It was wonderful, the enthusiasm in Brent's voice. Since the meeting in Barcelona and the merger, there had been so many anxiety-filled days and nights, but finally things seemed to be tilting back to a status quo. The new, supersized company was pushing forward on all cylinders.

Even the disaster with Neteller hadn't knocked them down—although it had been scary going for a while. Neteller had been a multibillion-dollar business, publicly traded, handling 80 percent of the processing market. When the two founders, John Lefebvre and Stephen Lawrence, were arrested simultaneously—Lawrence when entering the U.S. Virgin Islands—and charged in a sting operation headed by the U.S. attorney for the Southern District of New York, with help from the FBI, it essentially wiped Neteller from the map. Using UIGEA, the feds alleged that nearly 95 percent of Neteller's $7.5 billion in revenue came from online gambling. The Neteller case struck fear in the companies that had remained engaged in U.S. business—essentially, Scott's company, PokerStars, and Full Tilt, who together now had 90 percent of the market.

For a good week after that, Scott fully expected federal agents to burst through his door at any minute—even though his lawyers had assured him again and again that the fear was unfounded and ridiculous; that he'd done nothing illegal in Costa Rica, where he resided; and that even if someone decided to come after them under UIGEA, they'd have trouble building a case that implicated him personally, since the company was now owned by the Canadians, and he didn't handle the payment processing, and never had. Then again, the thought that Brent was at risk instead didn't give him any pleasure, but in his heart he truly believed that none of them should be afraid of an unfair law.

Still, the reality of the situation had affected him. Even his closest friends—Hilt and Garin—had commented on his growing paranoia.

Scott glanced at the gun in his lap while Brent went on about his most recent trip to visit processing agents and their financial offices around the globe. Brent's stories were pretty crazy. Since Neteller had gone down, the companies that had filled its place came and went almost daily, and were all shifty, sleazy affairs. They'd open, process deposits for a few days, then disappear, with whatever money they hadn't yet turned over. This happened again and again, yet still there was so much cash coming in that it didn't matter—a few hundred thousand dollars lost was almost immediately washed over by a few million dollars in player deposits coming in. Players never even knew what was happening behind the scenes—they could deposit and withdraw money just as before, without realizing where their money was going and how it was getting there.

But the shady processors and Neteller arrests were just a couple of components that were feeding Scott's paranoia. After UIGEA there had been quite a shake-up at their home office in Costa Rica as they absorbed the much larger Ultimate Bet. As with any merger, there were layoffs, but in a place like San José, layoffs led to death threats. Most had been things Scott could ignore: letters sent to the office, notes scrawled on cars in the parking lot. But a few of the threats had risen to a real, actionable level.

Getting a gun wasn't hard in a place like Costa Rica, where a hundred-dollar bill was enough to buy you just about anything you wanted and every taxi driver was his own little home shopping business, delivering drugs, girls, and firearms. But personal protection in a place like that didn't stop at a handgun; Scott had made some inquiries and eventually decided to hire a full-service security company. At that very moment, as he sat in his office listening to Brent, there were two bulletproof sedans parked in his driveway, and eight armed bodyguards in his living room. There was another standing outside his bedroom door, where his current girlfriend—Miranda, a fucking gorgeous *tica* with long, braided blond hair and a treacherously curvy body that could have graced the cover of a swimsuit catalog—was still sleeping off the night before.

He knew the number of guards was overkill, but he'd discovered that it was a real thrill going around town with a crew that large and imposing. In fact, after a week surrounded by an armed security contingent, Scott found he felt naked and vulnerable without them. Someone was always there, taking orders, making sure the road was clear; the door was held open; there was never a line for a bathroom. He, Hilt, and Pete would go to a restaurant or a club, and they'd show up in a caravan of armored cars, walk in surrounded by armed men—hell, by that point, they had more bodyguards than the president of Costa Rica.

The head of Scott's security company was ex-military, a guy named Santos, with a scar down the left side of his face and a very dark sense of humor. He talked about killing people as easily as he talked about the weather. Garin had taken a dislike to Santos from the start, telling Scott these weren't the sorts of people they were supposed to hang out with, that they weren't killers or mobsters, they were Internet entrepreneurs. But Scott reminded him that they were also expats in a country where life was held extremely cheap.

At some moments, though, Scott had to agree—Santos could be a little terrifying. When a laptop had gone missing from the office, Pete had tasked Santos with finding it. Not a day later Santos had brought the missing computer back. When Pete asked him how he'd found it, Santos had shrugged. He'd gone to a known neighborhood thief's house, kicked down the door, waved a gun, and asked where the computer was. Then he'd gone to another suspect's house and done the same thing. By the third door, someone had returned the missing computer.

Garin didn't want a guy like that hanging around, but to Scott, there was something undeniably thrilling about having a private army. And besides, with the amount of money they were generating, they were targets, whether Garin wanted to admit it or not.

"And things got really crazy when we got to Frankfurt," Brent continued, still going through his trip report. "Got a call just as we walked into the office, found out we needed to go to Munich to deal with another shit processor. The guy who runs Frankfurt says, 'Don't fly, take our cars.' He sends us down to the basement—and there were two black Porsches parked next to each other. Holy shit, man, these were nice cars. So we take the cars and we're doing Mach five all the way to fucking Munich."

Scott turned the .38 over in his lap, feeling the weight of the thing, the meaty heft of the grip. Porsches going Mach 5, armed bodyguards like his own private army to fix any problem, a couple million in revenue a day.

Maybe everyone was right—what did he have to be paranoid about?

"You sure this is a good idea?" Shay said from the front seat as Pete stepped out of the back of the taxi, followed by the Costa Rican girl in the tight jeans and white lace top. "We probably should have called first."

Pete shrugged, looking up at Scott's beautiful rented mansion on the hill. Most of the lights were still on, and there were two sedans parked out front, with matching tinted windows and armor-thickened doors and sidewalls.

"He never answers anyway," Pete said. "And besides, she says he's looking forward to seeing her. Who are we to argue with that?"

Shay laughed, getting out of the cab after him. The girl said something in Spanish, but Shay didn't translate, so Pete just smiled at her and nodded. He looked at the girl again as she bent to fix one of the straps of her bright yellow high heels. She was a pretty girl; her nails were a little too red and long, and her lipstick was almost blindingly bright, but she carried herself well for her height, which was all of five foot four. When they'd met her at the bar at Friday's and she'd asked if they could take her to visit Scott, they'd resisted at first. But she'd grown on them; she'd explained that she and Scott had dated a bit a few months earlier, and they still talked on the phone now and again—and she really wanted to see him. It was kind of funny—despite everything that had happened, some elements of Scott's personality hadn't changed at all from his SAE days. He was still the charming rogue, but now his playland seemed to have shifted from the University of Montana to the entire country of Costa Rica.

Pete headed up the steps to the front door, followed by Shay and the girl. It dawned on him as he reached the entrance that there was a good chance Scott already had a girl in the house; he wasn't in a serious relationship at that point, and there was hardly a week when there wasn't some gorgeous thing in his bed. But still, this one had been very insistent, and she seemed nice; maybe she had long-term potential. Since Pete, Garin, Brent, and Hilt all were in serious relationships now, it would have been nice to see Scott locked down as well. Maybe a good girl would calm some of his growing anxieties. They would all benefit from a less excitable Scott Tom.

Pete knocked on the door, and a meaty bodyguard by the name of Juan opened it. He smiled at Pete and Shay, then smiled wider when he saw the girl, and ushered them inside.

Scott was in the living room, sitting on the couch watching TV. There was a six-hundred-dollar bottle of wine in his left hand. He looked up, waved at Pete and Shay, and then he too saw the girl, and smiled as wide as the bodyguard. The girl ran over to him and gave him a big, wet kiss on the lips. Scott laughed, put an arm around her waist, and without another word led her out the side door to the back lawn, still clutching the wine bottle.

Pete looked at Shay, pleased with himself.

"I think we might have assisted in a love connection—" he started.

But before he'd even finished the sentence, there was a noise from the spiral stairway to his right, and he looked up to see another girl coming down the steps. She was tall, curvaceous, with long blond hair in tight braids hanging all the way down her back. She was dressed in a pink Juicy Couture sweatsuit, the front zipper down a few inches, revealing the soft curves of her ample chest.

Crap, Pete thought. He glanced at Shay, whose eyes were wide. The girl saw them and smiled amiably.

"Hey, guys," she said, in heavily accented English. "Where's Scott?"

Pete coughed. "I don't know. I think he went to get some beer."

It was a stupid response—there was probably enough alcohol in the place to satisfy an army. Before Pete could come up with another lie, the side door that led to the back lawn opened and the girl with the red nails strolled in, followed by Scott. The girl's shirt was completely unbuttoned, and she was in the process of adjusting her bra. Scott, for his part, was fastening his pants with one hand while taking a swig from the wine bottle with the other.

There was a frozen moment—and then suddenly the room seemed to split down the middle. The girl on the stairs came bounding down at full speed, her braids whipping out behind her. She caught the shorter girl by the throat and threw her to the ground, then leaped on top of her, screaming in Spanish. The shorter girl was screaming as well, trying to use those nails to defend herself.

A split second later, five huge bodyguards raced into the room, two from the kitchen, one from behind Pete and Shay, two from the backyard. It took all five of them to break the girls up. Finally, one of them got the tall girl over his shoulder and carried her back toward the stairs. Two of the guards grabbed the short girl by the arms and dragged her, still kicking and screaming, out toward one of the sedans.

Scott dropped to the couch, then put his head in his hands.

"And how was your night?" he mustered as the sedan screeched away.

CHAPTER 28

It was a Wednesday afternoon, sometime after two, and Pete was in a rush as he arrived back at the office. He wasn't in the habit of taking two-hour lunches, but there was just so much new marketing to go through he'd started using lunch as a way to meet with his affiliate reps. The way things were going, he would soon be using every hour of the day to cover all the ground he needed to. Too much business was never a bad thing, and in many ways he was in charge now—which was ironic, considering that he was the one who hadn't initially believed in the business, had never intended to join, and was most hesitant to continue after the UIGEA passed.

Yet here he was, rushing up the stairwell to get back to his desk. He had six phone calls to make before five, and two more after that, once the Korean software people came online. All of it had to do with managing marketing projects—a couple of new television shows on cable networks, a few promotions involving poker tournaments in cities around the world. If UIGEA had scared off the public companies on the London Stock Exchange, it had done nothing to curb the hunger of U.S. television networks. Although officially the online poker sites couldn't advertise money games, they could promote their free games, which easily linked into their real money games. AbsolutePoker.net was all free play, for fun—and that's the site that was promoted in the ads. But AbsolutePoker.com was for money, and that was where a good portion of the players eventually ended up. Because whatever Senator Frist and the good people in New York's U.S. Attorney's Office believed, people still wanted to play online poker—no matter who they had to give their credit card numbers to in order to get there.

Pete reached his floor and headed straight toward his cubicle. It was funny that he still had a cubicle; even though Scott and Hilt weren't coming to the office anymore and were no longer officially acting in any leadership capacity, nobody had yet claimed the one walled office on the floor. Perhaps it would always remain empty; it was kind of a metaphor for the business as a whole. Absolute Poker was a machine without a brain; all the gears still worked, the money still flowed in and out—but it was all automatic now, except for what Pete was doing in marketing and what Brent still did in payment processing. Joe Norton had brought in his own management team, including a CEO and a COO, who called the shots, but in many respects Pete and Brent were the new guard; the old guard had stepped away. Scott and Hilt still cared about the company, still wanted it to succeed. But they weren't living it anymore, day to day.

Pete had reached the entrance to his cubicle when he saw Brent approaching from the back of the office, followed by one of their in-house programmers, a squat fireplug of a guy named Angelo, a Costa Rican they'd recently hired to take the place of one of the Americans who had resigned shortly after UIGEA had passed. If Pete remembered correctly, Angelo worked part-time in Brent's old fraud-detection department; he checked the software for flaws and made sure the game play seemed kosher—looking for signs of player collusion, things like that. Every now and then the department found something it didn't like, and once in a while a player would get suspended. Nothing serious, but it was important that they monitored themselves as best they could, since there was no overarching regulation and, after UIGEA, it was unlikely there would be anytime soon.

Angelo was a head shorter than Brent, with thick glasses over sunken eyes, and he had one of those faces that always seemed to emit worry. Still, he was a good programmer, and even though he wore the Costa Rican unofficial uniform—shorts, T-shirt, and flip-flops—he was a solid worker.

"Cheer up, Angelo," Pete said by way of a greeting. "The rainy season will be over in four months. And then we only have the mosquitoes to look forward to."

It was a running joke between Pete and many of the Costa Ricans in the office; Brandi's dislike of the third-world country had become grist for a near-constant back-and-forth complaint session. Costa Rica was hot, dirty, lawless, and heavy on insects. The United States was uptight, moralistic, overregulated, and full of religious freaks who pushed Jesus like he was Colombian cocaine. But the look on Angelo's face seemed even more anxious than usual—he wasn't there to sling jokes.

"Pete," Brent said, his voice low. "There's something we need to talk about. It might be nothing—it's probably nothing—"

"What is it?"

Brent led Pete and Angelo into Pete's cubicle and ushered the programmer into Pete's seat. Angelo went to work on the keyboard, pulling up a website. Pete recognized it immediately as one of the more prominent poker blogs—a website called Two Plus Two. As a poker marketer, Pete knew the site well; the online poker community was pretty tight-knit and often rabid, and Pete had spent a lot of time reading through all the blog sites, although recently, he'd been too busy with all of the current promotional work to check in on them.

"What is it now? Someone complaining about our table felts again? Thinks they look too plush? Or is someone whining about not getting a withdrawal on time? If they knew the sleazy fucks we deal with since Neteller went down—"

"No, it's not that," Brent said. "Angelo?"

The Costa Rican began gesturing toward the screen. "This came to my attention a couple months ago, but I've been hoping it would just go away, as these things often do. Just people griping because they lost. But, well, it hasn't gone away; it's actually just getting bigger—"

"Spit it out," Pete said.

"There's been a bunch of players complaining on one of the High Limit tournament forums that three other players on Absolute Poker are winning a ridiculous amount, playing pretty suspiciously."

Pete cocked his head. "What do you mean, suspiciously?"

"Playing almost every hand pre-flop. Then, on the river, when they get bluffed, they seem to always call or raise. If their competition has good hands, they almost always fold. Like I said, a little suspicious. Anyway, after posting about it, a bunch of players analyzed the play history as much as they could, and now they've e-mailed us, asking us to take a look and see what the hell is going on."

Pete nodded. The truth was, this sort of thing happened all the time—accusations of cheating, either by other players or by the site itself. Usually the accusations were unfounded; everybody who'd ever lost in a casino believed, deep down, that the casino must be cheating. In games that employed some level of chance, strange things happened—and most of the time those strange things looked like cheating. But this sounded like a little more than common paranoia.

"And what did we find?"

Angelo shrugged. "Nothing yet. We're still analyzing all the hand history. It's going to take some time. But the idea that there's a superuser out there—"

He didn't need to finish the thought. Superuser accounts—accounts that could see all the cards as they were dealt, kind of like God mode on a video game—were the ultimate boogeyman of online poker; the idea that there were people on the inside, able to play so unfairly—it wasn't something anyone who worked in the industry ever wanted to even mention. Even the rumor of such an account existing could destroy an online poker company, because the whole industry was built on trust. If the players couldn't trust the online poker companies to keep the game fair, they were throwing their money away. If such a rumor spread on online forums, people would leave in droves. There would be a run on the bank.

There was a reason Absolute Poker had always been at the

top of the industry's rating lists: it took the security of its game very seriously.

"Send out a press release," Pete said, both to Brent and the engineer. "Explain that we've looked into this, that our investigation is ongoing—but that there is zero evidence that there's any cheating going on, or that anything like a superuser exists in our software. Nip this in the bud. A rumor like this could kill us."

"But what if—" Brent started. Pete shook his head.

"No what-ifs. Even a whiff of something like this can cost us millions. This has to end here. These guys are just jerking each other off, trying to find a reason why they lost money. Everyone gets paranoid when they lose. It's human nature."

Pete waited for them to file out of his cubicle, then headed toward his phone. He couldn't spend any more time on bullshit accusations from a poker forum. He had TV accounts to deal with.

Like he said, this had to end here.

A few days later Angelo and Brent were back in Pete's cubicle—and this time, both their expressions were equally grim. Not only had the situation not ended with the press release, it had gotten a whole lot worse. In fact, it had turned into something that might bring the entire company down.

"This can't be right," Pete was saying as he dropped into his office chair. He was only a few pages into the report Angelo had pulled together about what was going on, but already he could see—it was worse than anything he could have imagined.

"I'm afraid it is," Brent said. He looked at Angelo, who nodded.

"Does anybody else know about this yet?" Pete said.

"You mean apart from everybody on the poker blogs?"

Pete swallowed. His face was heating up, and he could feel the sweat beading on the back of his neck. "I mean Scott."

Brent shook his head. "I don't think so."

They'd have to bring it to him right away. Because if what he was looking at was true, even just the broad strokes, it was a disaster.

According to Angelo's report, the Absolute Poker press release had landed in the online forums like a lead balloon, but at first the complaints that continued to crop up all over the blogs were just that—complaints, without evidence, just more griping by losing players.

But then things took an unusual turn. A new player, who'd been hitting the AbsolutePoker.com tournaments under the handle CrazyMarco, had previously lost another tournament under what he considered to be suspicious terms. A player named Potripper had played almost all his hands pre-flop, then had played almost perfectly from there on—folding whenever he was up against a better hand, betting when he had the higher cards. It was almost as though he could see what everyone was getting dealt.

At the time, CrazyMarco had e-mailed customer service at Absolute Poker, asking for a play history for Potripper and the tournament. It turned out that, for whatever reason—either by mistake or on purpose—someone at Absolute Poker had e-mailed back an Excel file including the entire play history—everyone's cards, the history and information on everyone in the tournament's e-mail accounts. This was a staggering breach of protocol;

even worse, when CrazyMarco eventually looked through the file—spurred on by the Absolute Poker press release, which he felt had brushed aside a legitimate suspicion of cheating—he discovered that Potripper had indeed been cheating somehow. Playing the way he was playing simply by chance or skill would have been like winning the lottery ten times in a row. The only thing that explained his play was that somehow he could see everyone's cards, as they were dealt.

That was bad enough, but then things got really ugly.

Because upon analyzing the IP addresses and user details that had been provided by the anonymous Absolute Poker employee along with the hand history, it appeared that there was an observer account—number 363, to be exact—associated with Potripper's winning play—and that both 363 and Potripper's IPs could be traced back to Costa Rica. Once the blogger sleuths got hold of that information, it was just a few more steps, a little more research—and they'd uncovered the e-mail associated with account 363.

That e-mail was scott@rivieraltd.com. And according to the bloggers, that e-mail linked directly to the founder of Absolute Poker.

Scott Tom.

CHAPTER 29

"Eight hundred thousand dollars. He really screwed us. He really screwed *me*."

Pete felt like he was in the presence of a volcano that was seconds from erupting. A moment of tense silence swept through the room as everyone at the dining room table watched Scott struggling to regain control of his features. Pete had seen Scott mad before—even in college the guy could be volatile—but this was different. This was terrifying.

The moment had been building all through dinner. Even though the conversation had remained light, avoiding the obvious topic—the reason for the get-together in Scott's house—Pete could see the emotion roiling behind Scott's green eyes. By the time dessert was served, it was obvious he was using every ounce of willpower to keep that emotion from bursting forth until the dishes had been cleared and the wives and girlfriends had retired to the living room, the better to avoid getting hit in the crossfire.

Scott had been upset when Pete and Brent first told him about the discovery of the cheating scandal and where the players' intricate sleuthing had led them. Now that he'd had a chance to piece together what had happened himself—and had discovered how his name and reputation had been dragged into the depths of the scandal, without his knowledge—he was absolutely livid.

The e-mail had indeed been linked to his profile, though it was an e-mail he hadn't used for a long time. And the 363 account, it turned out, was in fact an old account also linked to him that had been deactivated years ago; it had been one of the employee accounts the Koreans had initially given them to keep tabs on the beta test, way back when they had first launched AbsolutePoker.com. Furthermore, when they traced the IPs of where number 363 and Potripper were being employed, it appeared that both accounts were being run through ports in computer networks linked through Absolute Poker's home servers: the cheating was coming from inside the house, so to speak.

Eventually Scott had been able to piece together what had happened. Sometime, perhaps as early as summer of 2007, a database programmer in Korea attempting to speed up back-end communication routines had inadvertently disabled the time delay in the software used to monitor game play. At some point after that an employee in Costa Rica had discovered this mistake—and had realized that with no time delay built in, it would be possible to see the cards as they were dealt, in real time. This employee had further realized that by using one of the old employee player accounts—number 363, to be specific—he could escape notice, because the old accounts often had large money balances. And since the account was linked to management, it wouldn't catch the attention of anyone in the Fraud Department, no matter how often it was being played.

From there, it hadn't taken long for Scott to discover the cheater himself. Scott hadn't been to the Absolute Poker offices for some time, and neither Pete nor Brent had made use of his computers since he'd stepped back from his official leadership role—but that didn't mean the computers had lain dormant. It turned out that one of the operation managers who'd worked part-time under Angelo had engaged account number 363. When he'd realized how easily he could beat the game, seeing all the cards as they were dealt, he'd gone to work with an unknown number of accomplices and had managed to swindle almost eight hundred thousand dollars from tournament players, observing the cards with account 363, then parroting that information to other handles—Potripper among them.

Worst of all, he'd done all this while inadvertently implicating Scott and the Absolute Poker headquarters.

It was appalling, and personally devastating to Scott, whose reputation was now being trashed on the poker blogs. The perception that he would somehow knowingly be involved in betraying the company he'd built—a company that was now bringing in two million dollars a day—by cheating his players was difficult to bear.

They'd immediately fired the person responsible, who besides being an operation manager was a close friend of Scott's—and had finally admitted to the public that they'd discovered the source of the scandal and had done their best to fix the software security issues. They had also responded by paying back anyone who had played against the cheating player accounts. But it wasn't enough, and they all knew it. Online poker was an unregulated industry; trust was something companies earned, and once players stopped trusting a company, they found somewhere else to play.

It was Pete who finally broke the silence at the table—surprising himself, because he couldn't even match Scott's angry glare.

"Let's look at this rationally. This is out now. It's been picked up by news organizations. The blogs, newspapers, it's gonna be on goddamn *60 Minutes*. We've already had a run on the bank. A week ago we were averaging two million dollars in revenue a day. Now that's been cut by more than fifteen percent. We're seeing withdrawals of around eight hundred thousand—a day."

The numbers were sobering. Pete had been monitoring the press and the blogs, and the issue felt like it was only growing, not going away. Absolute Poker was being linked in everybody's mind with cheating. This scandal was ruining the brand they had worked so hard to build.

"People are scared," Pete continued. He wasn't sugarcoating it, that was for sure. "The press is getting worse. At the end of the day, this is the largest online scandal in the history of gaming. It's a big deal."

"I know it's a big deal," Scott shot back. "And we're paying everyone back who lost. Hell, we're paying people back who would have lost anyway, even if this shithead hadn't cheated."

"That's not the point. It isn't really about how much was stolen. These guys sitting around playing poker twenty hours a day have nothing better to do than write and talk about this shit. This has become a soap opera. Ninety-nine percent of what everyone is saying is false—but we can't ignore the one percent that was true."

"We're getting goddamn death threats, Scott," Brent said, his voice so low it was almost a whisper.

Everyone looked at Scott, who finally laced his fingers together against the table, getting his anger in check. He spoke carefully, picking his words as if with tongs.

"I'm not even running things anymore. I've been gone awhile."

He seemed to resolve something internally; it was as if a shade was drawing shut behind his bright green eyes.

"Maybe it's time I made it official."

Scott understood that if they didn't get this scandal under control and behind them, it would cost the company much more than the money the cheater had stolen. But more important to Scott than the money was the company he had built, with his sweat and his blood and his passion—and he didn't want to watch it disintegrate.

He cared too much to ever let that happen.

The North American Sabreliner twin turbojet sputtered to life, then began rolling slowly toward the private strip of runway on the back lot of Santamaría International Airport. Up in the cockpit, the two pilots were finishing up their flight check, speaking in Spanish with each other and the tower, their voices carrying back through the open cabin of the midsize private business jet to where Scott, Hilt, Hilt's gorgeous blond wife, and Scott's girlfriend were opening a three-hundred-dollar bottle of champagne.

It was taking all four of them to attend to the bottle, because they were already a little buzzed, even though it was half past noon on a Monday. It had been a long weekend already, but they were now officially on vacation, taxiing toward the first hop on a weeklong Caribbean tour. The trip would begin with a refueling stop in Colombia, then extend through a handful of islands that promised white-sand beaches, palm trees, infinity pools, five-star hotels—and limited access to wireless networks.

It was a bittersweet moment for Scott, seated by the window in the back of the small plane, one hand on his girlfriend's arm, the other gripping a crystal champagne flute. Officially stepping down from his role at Absolute Poker had been more difficult than he'd ever let on to any of his friends; only his father knew the torment he had gone through as he'd made his leaving known to the shareholders and packed away whatever remained of his life at the company into cardboard boxes to be stored in the basement of his rented home. He'd never dreamed of his company growing so big—but he'd also never imagined that one day he would have to step aside, especially under such dark circumstances. Not just the cheating scandal caused by one of his employees, but what the UIGEA had wrought. Only his dad fully knew how painful it had been for Scott to watch himself go from being an Internet wunderkind, days away from being a billionaire, to having to separate himself from the company, forced to explain away his association with an entire industry that had overnight been tarnished by what he saw as an unfair and hypocritical act of Congress.

But here he was, in the back of a private jet with his closest confidant, who was struggling with the champagne cork as the plane bumped and jerked over the poorly paved runway.

The private plane had been Hilt's idea; none of them had

ever been in a private plane before, and it just seemed fitting. They were going out in style. At the moment, Scott didn't want to think about anything other than the trip to those white-sand beaches, those five-star hotels.

"Here we go!" Hilt shouted over the rumble of the twin turbojets churning to full power. They'd turned down the runway and were now gaining speed. There was a loud *pop*, startling Scott, but then he saw the cork pinging off the cabin ceiling and laughed. He tried to hold his glass steady as Hilt poured the champagne.

Then he turned forward, toasting himself as he watched the front of the plane lift upward, inch by inch, the front wheels coming off the runway, the nose tilting toward the sky.

And suddenly, a violent shudder reverberated down the right side of the cabin. Before any of them could react, the nose of the plane plunged back down to the runway—and there was an incredible noise, like a gun going off, as the tires blew. The pilots were screaming in Spanish, and Scott dropped his champagne glass, gripping the seat in front of him. He could see the pavement still flashing by through the front windshield. They had to be going 150 miles per hour, skidding on those trashed tires, the entire plane shaking and jagging like it was about to tear itself apart—and even worse, Scott could now see the end of that runway, a grassy field extending toward a fence and a grouping of what appeared to be steel pylons . . .

And then the plane hurtled off the pavement and into the field. A second later, they slammed into the first pylon; the windshield shattered inward, an entire section of the plane's nose shearing off. The cabin dipped down, and then they were spinning. One of the wings was caught in the grass, and jet fuel sprayed in through the cockpit, drenching the interior of the cabin. Part of the wing snapped off—and then everything went still. The plane was tilted halfway over, and there was a gaping hole where the cabin door used to be.

Scott's mind went blank as his reflexes took over. He grabbed his girlfriend in one hand, yanking her out of her seat belt. Then he was diving forward over the seats. Hilt and his wife were right behind him as he leaped through the opening and landed feet-first in a ditch, two feet deep in noxious jet fuel.

All four of them lost their footing as they struggled to crawl out of the ditch, but they didn't stop moving until they were a good three hundred yards from the wrecked plane. When Scott looked back, he saw the first fire truck pulling up. He dropped to his knees on the grass, watching with Hilt and the girls as the fire truck sprayed the crash site with foam from a giant, high-powered hose. They were all bleeding from scratches on their hands and faces, and they reeked of jet fuel—but somehow, they were alive.

By the time they were released from the hospital—more a precautionary stay than due to the severity of their cuts and bruises—word of the accident had already made it onto a variety of online local and international news sites and was rapidly spreading through the blogs. Scott's phone was gone, lost somewhere in the wreckage, which was now entirely presided over by agents from the FAA, since it had been an American-built airplane. They had to use Hilt's phone to check in with everyone to tell them that they were okay. By the second person they'd called, they'd realized that the story, spreading electronically at first, but eventually into newspapers as well, was turning into something out of a Hollywood thriller.

"This is ridiculous," Hilt said as he hung up the phone. "Now they're reporting that there was three million dollars and a bunch of coke in a suitcase in the back of the plane, and that you're on the run to Colombia. I've never gone near cocaine—and where the hell did they get the three million dollars?"

Scott shook his head, bewildered. He was watching an urban legend generating right in front of him, and there was nothing he could do about it. What the hell—it was just too perfect to fight. A high-flying American cowboy from Montana, founder of an online poker empire, fleeing Costa Rica to Colombia with a suitcase filled with millions of dollars and mountains of coke.

"If you've got to go out," he said finally, laughing, pulling at one of the bandages on his hands, "may as well go out with a bang."

CHAPTER 30

If Pete had thought things would return to normal after Scott exited in his faux blaze of glory, he couldn't have been more wrong. The Absolute Poker cheating scandal that the bloggers had uncovered, as bad as it was, was only the tip of a much, much larger iceberg. This time, however, they were facing a problem they hadn't created, but had inherited—or, more accurately, had bought.

The whispers had started right after Barcelona, but everyone had been too busy dealing with the massive jump in their business and the payment-processing fallout of UIGEA to take any real notice. Just as with the Absolute Poker scandal, the whispers were coming from players, via poker blogs; but this time, as soon as the rumors reached Pete in his cubicle in San José, he took them extremely seriously and reacted as quickly as he could. He'd learned his lesson—covering up suspicions of scandal only made the scandal worse—and this time the scandal appeared to be so much bigger from the outset that no number of press releases would make it go away. It would take months to get to the core of what had happened, but once Pete had his proof, he intended to act, and to put everything out in the open, as clearly as he could.

When the time finally came, he turned the office that Scott no longer used in their San José headquarters into a sort of war room. The walls were covered in charts and graphs, most culled from the blog sites, a few developed by his own in-house software guys. He'd set up a small round table in the center of the room, which he, Brent, and Angelo were now seated around. Computer printouts, faxes, and studies sent from Korea were piled high around them; many of the detailed reports that went along with the studies were actually in Korean, but Pete had gone through enough of it over the phone with C.J. that he knew the gist of what they had found.

It had taken thirty people, and more than a million dollars in investigative fees, to build the evidence on the table in front of them. And Pete was now certain; just as many bloggers and players had been posting over the previous six months, there was a continuing and severe cheating scandal taking place at the tables of UltimateBet.com. The Absolute Poker cheating scandal had been relatively short-lived and had cost players between six and eight hundred thousand dollars. The Ultimate Bet scandal, according to Pete's research, had possibly gone on for years, dating all the way back to at least 2005, and had probably resulted in many millions of dollars stolen from players—perhaps between ten and twenty million, if not more.

Even worse, Pete and his investigators believed that the cheating went all the way up to the top levels of the Ultimate Bet corporate brass. Pulling IP addresses had given them names and profiles way up in the hierarchy of their former competitor. Once he'd compiled the evidence, Pete had confronted UB's leadership directly; he'd shown them his evidence of back-door programs, irregular play, chip dumping—all of it. And they'd just dug in their heels. But there was no way Pete could let this go.

It was too staggering a find. It was only a matter of time before it became public, because the players analyzing the suspicious play of many Ultimate Bet accounts were uncovering more evidence of cheating every day. Pete also intended to put out a press release about what he'd found, and to get the Kahnawake Gaming Commission involved. The news was going to seriously impact the business—and the value of the entire company.

From what Pete could see, all of this had been going on before they bought Ultimate Bet during the Barcelona conference—and had continued since.

"They fucked us," Pete said, as Brent and Angelo leafed through the pages for the hundredth time. "Whether they did it knowingly or not, they fucked us big-time. And they're still holding that promissory note for around a hundred and thirty million dollars. We're supposed to pay them a hundred and thirty million bucks for a company with this shit at its core."

Brent's shoulders sagged. "This is going to cost us. But I'm not sure what we can do. We can't go after them in court. I mean, what court? What jurisdiction? Certainly not the U.S. How do we deal with this, legally?"

Pete shook his head. He had given this a lot of thought. "I don't think we deal with this in court, at least not initially. I think we handle this old-school."

Brent looked at him. Angelo started to fidget in his chair.

"What do you mean, old-school?"

Pete grinned. "You remember Hell Week, back at SAE?"

"Of course."

"I think it's time we bring Hell Week to Central America."

And then Brent was grinning too. Angelo looked at both of them, then forced a grin as well, although he had no idea what the hell they were talking about. Angelo couldn't possibly understand—but those cheating fuckers at Ultimate Bet were about to get a taste of all-American fraternity life, SAE-style.

CHAPTER 31

The limousine was pitch-black, its bulletproof windows tinted so dark they were like glassy bat eyes, flashing intermittently as the bright slice of moon blinked down through gaps in the thick jungle overhang. The road was so narrow, the curves so steep, that at times the long, sleek car seemed to be tilting almost ninety degrees from horizontal; the four overweight men jammed together in the leather-lined, sectioned-off backseat were sweating through their tailored suits, emitting gasps of fear whenever the car tilted a little too precariously toward a steep ravine or came a little too close to a jagged rock face.

The men weren't sweating just because of the treacherous road. The air-conditioning in the car had been turned off, which was particularly torturous because they were presently deep in the hills outside of San José, and it was weeks into the region's humid season. Not that Costa Rica had a season that wasn't humid, but these men were not used to the tropics, or the third world. They had flown in from London, Vancouver, and Portland. They were businessmen, middle-aged, two with law degrees. And they had fully expected a modern hotel, a glass-walled conference room, maybe a tray filled with Starbucks in Styrofoam cups.

Instead, they had been met at the airport by three burly security men, wearing obviously visible sidearms right out in the open. They had been ushered into the back of the limo and driven directly into the jungle.

Two hours had gone by like that, the limo twisting up and down the narrow road, and the men were about ready to break down. Thankfully, it was only another ten minutes before the car finally slowed to a stop, angling into a dirt clearing in front of what appeared to be a single-story wooden shed. Through the heavily tinted windows, it was hard to make out much about the decrepit building, but there were obvious holes in the thatched roof, and there was a pair of mangy-looking dogs tied up near what appeared to be an outhouse next door.

"Christ," one of the men uttered. The three armed security guards didn't respond. One remained behind the wheel, the engine running, while the other two came around the car and opened the passenger doors.

The suited men stumbled out into the thick jungle air, stretching their legs. One of them looked like he was going to throw up, but he managed to contain himself. Then a security guard pulled a long metal wand out of his jacket pocket.

"Strip down to your underwear," he said gruffly.

One of the men laughed. The guard stared at him with narrow eyes. He didn't put his hand on his holster, but his fingers twitched in that direction. The businessman stopped laughing, then glanced at one of his colleagues.

"Are you serious?" the businessman asked. "Absolutely not. We will do no such thing."

"You strip, or there's no meeting. And you can find your own way back to the airport."

The second security guard began to head back toward the car. The businessmen looked at one another. One of them cursed.

Then, slowly, they started to undress.

It took a full five minutes for the heavyset men to get down to their underwear, each suit piled up on the dirt in front of them. They stood there with pasty flesh, boxer shorts and white Hanes briefs, beads of sweat rolling like marbles down trembling legs.

The guard stepped forward and began waving the wand over each man, then over the piles of clothes. None of the men noticed that the wand was actually a television remote control affixed to a car antenna, or that there were no buttons, lights, or batteries involved. They just waited, terrified and sweating, for him to finish.

When he was done, he gestured for them to head into the shed. The men looked from him to their clothes.

"Like this?" one of them asked. "Can't we get dressed?"

The guard shook his head. "The boss says you guys go in like this."

Cursing even louder, the four businessmen, still in their underwear, hurried toward the shed door.

It wasn't until the men were back in the car, still buttoning and zipping up their clothes as they headed back down the jungle road, that Pete and his negotiating team, still in the dank, stark confines of the shed, which reeked of goat dung and rotting produce, let themselves burst into hysterical laughter. Pete actually fell to the dirt floor, he was laughing so hard, and it was a good few minutes before he finally caught his breath.

The plan had worked perfectly. Those businessmen were worth tens of millions of dollars, were at the top echelons of what was formerly a publicly traded company—and there they were, in their freaking underwear, agreeing to anything and everything Pete and his team put forward. Pete had accused them of stealing $25 million, then demanded that they cancel the $130 million debt, in exchange for a simple $1.5 million payment. Furthermore, he told them he fully expected a renegotiation of their profit-sharing ratio, and that he intended to pay back whatever he could that had been stolen from players over the past half decade.

The businessmen had agreed to every last demand. Pete couldn't help but feel proud of himself. He wasn't just a master at marketing; if his time as SAE president had taught him anything, it was how to pull off a damn good fraternity hazing.

CHAPTER 32

APRIL 15, 2011

It was a few minutes after 9 A.M. on a Friday, one of those perfect April mornings that makes living in Costa Rica worth it, despite all the negatives. Any argument Brandi could make about third-world conditions, the traffic or the smog or the power outages—none of it had any resonance on a morning like this, against that sun, that breeze, the smiles on every face in the office. Pete was the only one who seemed to be working at his desk in his cubicle, while everyone else was already making plans for the beach; but really, he was thinking more about playing golf that afternoon with Brent than he was about poker revenues and television buys. His mind was already on a golf course, his eyes following an imaginary white ball as it arced across the aquamarine tropical sky. He didn't even hear the phone ringing on his desk until an accountant who happened to be walking by, tapped his shoulder and pointed at the receiver.

The voice on the other end of the line barely cut into his daydream. It was one of his advertising contacts, a guy who was part owner of a poker magazine.

"Pete, I think you're getting hacked."

The man's voice sounded pretty anxious, but Pete knew it wouldn't be more than a minor irritation. Over the previous few months, as the art of hacking had become more in vogue, they'd dealt with a handful of hack attempts, and usually it was just some kid in Russia or China taking a shot at their software for no reason other than boredom.

"Thanks for the heads-up," Pete responded, watching that spinning golf ball in his head. "I'll get someone to take a look at it—"

"I think you better check it out for yourself," the magazine owner said, and then he abruptly hung up.

Pete sat up in his chair. He was still sure it was nothing—the business had been going so well since they'd dealt with the Ultimate Bet scandal and resumed building their player base until it was almost as high as it had ever been—but the tone of his contact's voice had gone from anxious to something far less identifiable. He put the receiver back on its base, then powered up his computer. A second later he typed in the address for AbsolutePoker.com.

And his heart nearly stopped in his chest. His entire screen was taken up by the official seal of the United States Department of Justice. Pete stared at the big eagle holding the branch, at the official-looking titles, at the red border, and then he started to comprehend the words.

This site has been seized . . .

Pete's first thought was, *Shit, this is a serious hack*. And then a much darker thought crossed his mind. He quickly typed in the domain name for PokerStars, their main competitor who had stayed in the U.S. market. A second later he was staring at the same DOJ seal, the same horrific words:

This site has been seized . . .

He quickly typed in the address for Full Tilt Poker, the other major poker site that had U.S. customers.

This site has been seized . . .

His next thought was, *What are the chances that the three biggest poker websites in the U.S. are getting hacked at the same time?* And then he closed his eyes. He knew it was impossible. It hit him right then.

This was real.

This was *it.*

"Everyone!" he shouted, his voice reverberating off the cubicle walls. "Get on your computers. Go to the site."

There was a pause, and then a handful of gasps. The sound ran like an infection from cubicle to cubicle, as everyone else in the office went to the site as well. Then he heard someone cursing—followed by the sound of a phone crashing into a wall. Without thinking, Pete brought his hand up in the air, turned it into a fist. And before he could even comprehend what he was doing, he crashed it down against his keyboard, again and again. The plastic shattered, keys raining across his desk.

Through all of the scandals, the cheating, the RCMP raiding Vancouver, everything, it had never really been *it.* Even before he had been with the company—the Caribbean bank failures, Shane's addiction, Scott's car crash, the UIGEA—it had never been *it.*

But in that split second, Pete knew that it was over.

After Vancouver, his wife had asked him, *When do you quit?* And he'd answered, *When it's no longer a gray area. That's when we get out.*

In that moment, Pete knew—it was no longer gray. It was black-and-white, and it was time to get out.

Brent had one hand on the steering wheel, the other tapping at the controls of the MP3 player built into his dash. He was trying to find just the right music for the short drive to the office. He was a little late, but he didn't think anyone would care. Then again, he was pretty much the boss now—well, him, Pete, and a few others. They'd been sharing responsibilities since Scott and Hilt had left, and everything was going so well, nobody would have even cared if he'd taken the entire day off—turn the afternoon of golf he'd planned with his friends into a three-day weekend. Golf, beach, hanging out with his wife and two kids—Christ, now it was *two* kids, two amazing little balls of energy, one five, the other a little over one—crazy to even think about.

And then his cell phone was ringing. He abandoned the MP3 controls and put the speaker to his ear.

To his surprise, it was his lawyer.

"Brent, you need to get to the office right away, and call me as soon as you get there."

Brent raised his eyebrows, because he'd never heard his lawyer sound like that.

"Why? What is it? Did someone die?"

"We just got a fax from New York. It's an indictment. And your name is on it."

It felt like the universe was crashing down around him. He needed to get off the road, immediately. He was still a good ten minutes from the office, but the beach traffic was snarling in front of him. He yanked the wheel to the right, driving into the parking lot of a gas station, then slammed on the brake.

"What do you mean, indictment? Who's indicting who?"

"The U.S. Attorney's Office for the Southern District of New York, with the help of the whole DOJ. They've shut down all the sites, and they're bringing up ten of the principals on charges."

"Wait, what? What charges?" Brent's voice was little more than a croak, echoing around the car, blending with the noise from the traffic crawling past the gas station. "Which principals?"

"The charges range from operating an illegal gambling business to bank fraud to money laundering."

Money laundering. Bank fraud. The words reverberated in Brent's head. This wasn't about whether poker was a game of skill or risk. These were real, criminal charges. These were the kind of charges they threw at actual criminals.

"Who does the indictment name?"

"It looks like they picked two people from each of the big three poker companies that continued to operate in the U.S. after UIGEA, and a payment processor. Two from PokerStars, two from Full Tilt, and two from Absolute Poker. From Absolute Poker, they picked you—and your brother."

Brent lowered his head to the steering wheel.

"It gets worse," the lawyer continued, his voice seeming so damn far away. "They've seized all the company funds they could get their hands on. Seized all the domain names. The payment processors—it looks like they're just disappearing, a lot of them taking players' money with them . . ."

Brent was barely listening. His eyes were welling up with tears.

Money laundering.

Bank fraud.

"This is what it looks like," he whispered to himself, "when everything comes crashing down . . ."

Some people just knew how to live.

Scott put his feet up on the wooden banister, staring out over the white-sand beach, watching the soft waves lapping at the seashells. The carved-driftwood balcony of the small, elegant restaurant at the five-star hotel where he was staying was crowded—mostly well-dressed tourists finishing up breakfast, planning their day on an island that could only be properly described as a true paradise—but Scott had found a nice, quiet corner in which to relax by himself. His girlfriend was still up in the room, sleeping off a night of dancing in the hotel bar. Soon Scott would rejoin her, and they would hit the beach—or not; there was a nice big Jacuzzi in their suite, overlooking those same soft waves. Hell, he thought to himself, he could stay in that suite all day. He could stay on this beautiful island forever.

And then he felt a trembling against his thigh—his cell phone, jammed into the side pocket of his bathing suit, on vibrate mode. He thought about ignoring it, but decided it could only be good news. It was that kind of a morning.

He got the phone free and placed it against his ear. To his surprise, it was Hilt on the other end of the line. As far as he knew, Hilt was in Panama, setting up his new home. Now that they were no longer involved in the company, they had gone off in different directions. Hilt had landed in Panama City and was working on new business ventures, using that fierce brain of his to begin his empire-building anew. Scott was still searching, but the wonderful island of Antigua had seemed as good a place as any to start.

"Hey, buddy," Scott said, watching a seagull hop just out of reach of the waterline, bouncing on legs the thickness of twigs. "You should get your ass over here, this place is awesome—"

"Scott, stop and listen. The site was just shut down by the Department of Justice. The U.S. Attorney in Manhattan filed an indictment against you and Brent."

Scott laughed. "April Fools', right? You're a bit late, fucker. It's the fifteenth—"

"This isn't a joke. I'm e-mailing you the indictment right now. It's gonna take you an hour to read through the whole thing, but they're trying to get you on running an illegal gambling venture, bank fraud, money laundering, and a bunch of other shit. It's coming from New York, because of their broad gambling laws—but Scott, this is a real indictment, with real jail time at the end. The lawyers I've talked to said you could be facing a lot of years—"

"Wait, what?"

"Scott, listen to me! You and Brent were just indicted."

Scott blinked. This couldn't be happening now. So many years had passed, so many hurdles had been overcome—why was this happening now?

"Why *me*? I'm not even there anymore. Brent's been dealing with those fucking processors—but why is my name there?"

Hilt didn't answer. The blood was rushing through Scott's head. He couldn't believe what Hilt was saying. He'd never been charged with anything more than a traffic ticket in his life. He hadn't done anything wrong. He'd lived pretty hard, he'd gone through some crazy shit—but in his mind, he hadn't done anything *wrong*.

"This is insane. Hilt, what should I do?"

Hilt still didn't say anything. Scott knew what his friend was thinking. It could have been any one of them named in there. It felt so arbitrary and unfair. The U.S. government could have addressed the business in so many other ways. It could have come up with regulation, it could have demanded taxes, it could have even filed a cease and desist. But it had gone right for the jugular.

Still holding the phone, Scott stood up from his chair. He found himself stumbling back from the banister, toward the small tiki bar in the corner of the restaurant's balcony. When he got there, he ignored the smiling bartender and pointed toward a bottle of tequila on a low counter behind the man.

As the man poured him a shot, he spoke low, into the phone.

"I'm gonna fight this," he said, trying to sound hard. But inside he was falling, spiraling down toward a deep hole that had been with him his whole life. He'd climbed so far out of that pit—but now he was plunging right back down.

"Scott," Hilt was saying, "you need to talk to the lawyers. This is going to be big news. The newspapers are already calling it Black Friday. Maybe you can go to New York and make some sort of deal."

Scott hung up the phone. He knew Hilt had his best interests in mind, but Scott wasn't going to New York. He was going to stay right where he was, in Antigua, as long as he could. He truly believed, in his heart, that he was innocent, and he wasn't going to throw himself into hell because of an indictment that had come out of an unfair, poorly written law.

He had built a company out of an American pastime. He'd created a way for the Internet to provide a game of skill to people who wanted to play it, and now he was being persecuted for it. He'd come from nowhere, and had built a billion-dollar company—and now it was all gone, a multibillion-dollar industry destroyed in the stroke of a pen, and he was facing jail time.

No, he was going to stay right where he was.

CHAPTER 33

Lawyers, lawyers, and more lawyers. Phone calls, e-mails, faxes. Long hours sitting in glass-walled offices, staring at documents that barely made sense, poring over numbers that never added up. Day after day, week after week, and yet still it just kept spinning on and on, a never-ending marathon, so mind-numbing and heavy that by the end, when Brent finally made his decision to turn himself in, he was actually relieved.

Of course, the decision hadn't been easy. And at first the thought of what he, personally, was going to do had taken a backseat to how he and the company would deal with the fallout of what everyone now called Black Friday.

An entire multibillion-dollar industry, gone in a single day—the fallout was astronomical, and almost instantaneous. The minute those three websites were seized, their assets frozen, hundreds of millions of dollars in players' accounts were suddenly put into jeopardy, and much of those millions disappeared in the blink of an eye. Likewise, an entire marketing and advertising field vanished—perhaps a hundred million dollars in television sponsorships alone. Overall, throughout all the industries affected, perhaps as much as ten billion dollars evaporated. But to Brent, and to many millions of people around the globe, the biggest disaster fell on the players themselves, many of whom were facing personally devastating losses.

Though PokerStars had enough capital on hand to pay back many of its accounts, Full Tilt and especially Absolute Poker were facing true catastrophe.

Brent saw it happening right in front of his eyes. Almost immediately a large number of his payment processors—shady middlemen in the best of times—simply disappeared, along with whatever money they owed players and the company. At that same moment, Brent and Pete were also facing the forced firing of more than four hundred employees, most of them in Costa Rica, all of them demanding severance, whatever salaries they were still owed, whatever benefits the laws of their country mandated. And on top of all that, of course, there were the lawyers—not just his, but the company's, the shareholders', the employees'. Everyone now needed a lawyer.

From the very first moment that the site was frozen, the money began to evaporate—faster and faster, and it looked like it was only going to get worse.

The first thing Brent tried to do was to submit a proposal to the U.S. government to help the players reclaim at least a portion of the money they'd deposited into the site to play poker. By his calculations, the company had about fifteen million dollars in cash, another forty-five million in receivables. His plan gave the players precedence over other debts—and, by his numbers, would have gotten them about seventy-five cents on every dollar they'd deposited.

Unfortunately, the government turned his plan down, for a variety of reasons. They were in the process of building their case, and the most important thing to them was their investigation, and taking down what they considered an illegal industry. The lengths they'd already gone to were impressive. Brent learned from his lawyers that beginning back in 2009, the Department of Homeland Security had created a fake payment processor, called Linwood Payment Solutions, which was really staffed by U.S. federal agents. Linwood had actually processed close to fifty million dollars in poker money, while the feds gathered evidence against Brent and the others named in the indictment. Brent himself had dealt with Linwood—signed his name on forms that were now entered into evidence against him.

Clearly, the government was not interested in Brent being involved in cleaning up the fallout that Black Friday had caused; the only thing they wanted from him was cooperation—in the case against his company and the industry as a whole.

And with two young kids at home and a Colombian wife, whom he desperately wanted to bring back to the States to try to restart a normal life, Brent had taken the government's request for cooperation extremely seriously—until his lawyers explained what "cooperation" really meant.

It wasn't a matter of telling them a few stories and giving them details about how the business worked. If he cooperated, he would have to speak out against anyone the government wanted to target. His friends, his fraternity brothers—and, of course, Scott. There was even talk of Brent having to wear a wire.

In exchange, Brent would get a guarantee of no jail time and no forfeiture of assets. But even so, he simply couldn't do it. His brother was his hero, and despite everything that had happened, his brother would always be his hero.

The brief conversation he'd had with Scott right after the indictment hadn't changed anything—although it had been incredibly tense and had started on a very harsh note. The minute Scott had gotten on the phone he was shouting in Brent's ear:

"Why is my name on this fucking thing?"

But Brent's response had brought him back into the reality of the moment. "Why are you asking me? My name is there too."

And after that, it had just been two brothers talking about a future they'd almost had, a past they'd shared, and what all of them would do next. Scott was going to stay in Antigua. He'd probably end up trapped there, a fugitive; Antigua wouldn't extradite him, because there, running an online poker site wasn't illegal, it was a respected line of work. Pete had gotten on the first plane to the United States, taking his family with him. Garin was remaining in Costa Rica—even though at the moment it seemed a little terrifying. Right after the indictment, Scott's house was raided by Costa Rican agents, who'd literally carried out everything of value they could find. Hilt was in Panama; technically, he was free to go where he wanted, but he was going to avoid setting foot in the United States, because the last thing he wanted to face was a subpoena. None of them wanted to be forced to speak out against one another.

And that left Brent. He had a wife, two young children, and no job, and he was facing a near future of lawyers, lawyers, and lawyers. All he wanted was for all of it to be over.

So he'd made a decision—not to cooperate fully, but to turn himself in.

Over the next few weeks his lawyers had begun to negotiate his surrender with the U.S. government and the U.S. Attorney's Office. The government had quickly agreed to drop most of the gambling and money-laundering charges, but Brent would have to plead guilty to some form of bank fraud. Even so, the prosecutors were willing to admit that it was fraud without any loss to anyone—without any real victims. In fact, the people being defrauded were making money, not losing it, and most were willing participants.

Although the sentencing guidelines for what he would plead guilty to were between twelve and eighteen months in federal prison, his lawyers, and the prosecutors, were confident that his past, and his willingness to surrender, would translate to a sentence of probation. Maybe he'd have to pay a fine, but he'd be home, he'd be free, he'd be with his family.

In the end, the plea had seemed the right thing to do.

On July 21, 2012, even though the Department of Justice's own attorneys argued for leniency, Brent Beckley was sentenced to fourteen months in prison for his role in running an online poker website. The very prosecutors arguing the case had agreed that there was no evidence that Brent had caused losses to any banks or any individuals. Despite this, the judge enacted a sentence that, in his words, "makes clear that the government of the United States means business in these types of cases."

After the sentence was handed down, Brent found himself standing in the back of the courtroom, surrounded by eight members of his extended family. Everyone around him was crying, and Brent himself was in a state of near shock. Even so, he did his best to stay upbeat.

"It's not cancer," he said as he gave them each a hug, one by one. "We'll get through this."

Even through his shock and sadness, there was a sense of relief. Because for him, at least, it really was over.

EPILOGUE: THE AFTERMATH

Brent Beckley is currently serving fourteen months in a minimum-security federal prison situated on the grounds of a supermax outside of Denver, Colorado. He is taking culinary classes while in prison and hopes to work in the restaurant industry upon his release.

Garin Gustafson and Shane Blackford are both happily married, working together as business partners and entrepreneurs, and are still living in Costa Rica. Garin visits Montana frequently, and Shane has a new baby daughter and is still actively involved in his NA and AA fellowships.

Pete Barovich finally listened to his wife and got his family on the first plane out of San José after the federal indictment. He has since relocated to Phoenix, Arizona, where he runs his own business.

Hilt Tatum remains in Panama City. Since leaving Absolute Poker he has started various businesses and currently works in private equity while commuting back and forth to the UK to finish up business school at Oxford University.

Scott Tom remains trapped on the island of Antigua; although he is not technically a fugitive—since he was on Antigua when the federal indictment came down, and thus isn't *fleeing*, but remaining still—the feds currently consider him "at large." Despite the fact that he presumably earned millions of dollars as the creative force behind Absolute Poker, and was once days away from becoming an Internet billionaire, he now lives in what could be accurately described as a gilded cage.

The industry of online poker took a huge hit on Black Friday; billions of dollars vanished with the stroke of a pen, and millions of players saw their poker accounts disappear. Even so, most legal experts believe the U.S. online poker business will one day make a regulated, taxed, and extremely profitable recovery. In July 2012, PokerStars agreed to pay a $731 million fine to the U.S. government; part of that settlement included a deal in which PokerStars purchased what remained of Full Tilt Poker. The newly merged company has since resumed operations; in January 2013 it announced that it had purchased a brick-and-mortar casino in Atlantic City, gaining a foothold back in the American market. Even more significant, one month earlier, in December 2012, the Department of Justice's Office of Legal Counsel finally and officially clarified its position on the 1961 Wire Act as it applies to the online gaming business. According to U.S. deputy attorney general James Cole, the OLC analyzed the scope of the Wire Act and concluded that it is indeed limited to sports betting. In other words, the DOJ now officially believes that online poker should not face criminal prosecution under the Wire Act.

Whether these developments open the door to the resurrection of the U.S. online poker industry remains to be seen.

PHOTO APPENDIX

March 2002: Group shot on the scouting trip to Costa Rica. Garin Gustafson, Gary Thompson, Phil Tom, Shane Blackford, and Scott Tom. *(Courtesy of Garin Gustafson)*

September 2002: First trip to meet with Korean software programmers—a partnership begins. C. J. Lee, Garin, and Victor Kim. *(Courtesy of Garin Gustafson)*

June 2006: Garin at the Gumball 3000 finale party in the Playboy Mansion. *(Courtesy of Garin Gustafson)*

Absolute Poker software in the beta testing period, depicting the initial design of a virtual poker table.

The foam-covered aftermath of the crash of the private plane, from which Scott and Hilt somehow emerged alive. *(Courtesy of Oscar Hilt Tatum IV)*

ACKNOWLEDGMENTS

First and foremost, I am grateful to the many individuals within this story, named and unnamed, who opened up their lives to me over the many months it took to research this amazing tale. I am also indebted to my wonderful editor, Peter Hubbard, and the team at William Morrow. I am also, as usual, grateful to Eric Simonoff and Matt Snyder, the best agents in the business. Many thanks to my Hollywood brother Dana Brunetti and the incomparable Kevin Spacey, as well as Mike De Luca. I would also like to thank my secret weapon, Jeff Glassman, and his associate Michael D'Isola. Many thanks to Barry Rosenberg, Megann Cassidy, my incredibly supportive parents, and my brothers and their families.

And most important of all, thank you Tonya, Asher, Bugsy, and our newest addition, Arya—you make it all worthwhile.

BOOKS BY BEN MEZRICH

STRAIGHT FLUSH

The True Story of Six College Friends Who Dealt Their Way to a Billion-Dollar Online Poker Empire—and How It All Came Crashing Down. . .

Available in Paperback and eBook

"A breakneck retelling of how a bunch of closely knit fraternity brothers built an online poker empire . . . Fast-paced and wild."
—*Kirkus*

RIGGED

The True Story of an Ivy League Kid Who Changed the World of Oil, from Wall Street to Dubai

Available in Paperback and eBook

"Gripping and stylish, *Rigged* keeps the reader moving through its pages at the pace of its characters' Ferraris and luxury jets."
—Matthew Pearl, author of *The Dante Club*

BUSTING VEGAS

A True Story of Monumental Excess, Sex, Love, Violence, and Beating the Odds

Available in Paperback and eBook

"Mezrich skillfully blends the sugar of the high-roller lifestyle with the medicine of the mathematics behind 'the three techniques' to create a concoction that goes down with entertaining ease." —*Boston Globe*

UGLY AMERICANS

The True Story of the Ivy League Cowboys Who Raided the Asian Markets for Millions

Available in Paperback and eBook

"A high-octane passion play pitting a young man's ambition against his sense of humanity." —*Oregonian*

www.benmezrich.com **www.myspace.com/benmezrich**

Visit HarperCollins.com for more information about your favorite HarperCollins books and authors.

Available wherever books are sold.